Will Climate Change Your Life

How to drive a 4x4 and still save the planet

Anthony Day

Will Climate Change Your Life?

© **Anthony Day 2007**

Edited by **Lesley Morrissey**
and the team at *www.thebookcooks.com*

Cover Design and Illustration by **Ellie Smith**
www.ellie-draws.co.uk

Book Design & Typesetting by **Martin Coote**
martincoote@googlemail.com

Set In *Optima* 11 on 14pt

First published in 2007 by;

Ecademy Press

6 Woodland Rise, Penryn, Cornwall UK. TR10 8QD
info@ecademy-press.com • www.ecademy-press.com

Printed and Bound by;

Lightning Source in the UK and USA

Printed on acid-free paper from managed forests. This book is printed on demand, so no copies will be remaindered or pulped.

ISBN-9-781905-823109

"Gripping, knowledgeable explanation of a huge subject, cleverly breaking it into manageable slices for all to take note and consider actions. Ignore the warnings of this book at mankind's peril."

Joe Morris,
MD, Waterman Sustainable Energy

"A comprehensive, plain English synoptic summary of the technological, economic and socio-political options we need to consider as our species moves into the uncharted territory of decoupling species evolution and carbon use. Full of useful comparator data/reference material for the beginner and the seasoned realist in this area."

Peter Jones,
Director of Development and External Relations, Biffa Ltd

"Analysis and assessment of the factors that may be influencing climate change is a complex business. This book provides a useful, accessible, well researched and balanced assessment of current viewpoints and evidence. It also analyses potential future energy sources as well as providing a practical assessment of how to most cost-effectively reduce your own impacts on carbon emissions."

David Turley,
Defra's Central Science Laboratory

"While the causes and timescales of climate change are fairly well understood, the world is in disarray about the solutions. Here is a book that takes the biggest and most difficult topic or our time and explains in simple terms what it means for the UK and what we should be doing (and not doing) about energy supply and demand to avoid serious disruption to our way of life."

Michael Green,
Director, the UCG Partnership.

"Big Green Book highly recommends this enjoyable book to everyone wishing to gain a broader understanding, of the issues over Climate Change. The shifting sands of climate politics have now put tackling climate change firmly on the agenda, and taking action to combat Climate Change is now becoming a global economic necessity. This book speaks to both the individual and company employee, who want this world to prosper in perpetuity. It helps bridge the gap in today's knowledge and concerns surrounding Climate Change on both today and tomorrow's economies and the lower carbon future that will need to be achieved by us all."

Stephen Lings

"This book is about putting some realistic perspective into the discussion on global warming, and is well worth reading, especially by anyone confused by the hype they have read in the press.

Written in an approachable style, this book is aimed squarely at the average man, or woman, in the street and does an excellent job of presenting a series of highly complex arguments in a fair and balanced way. With this book you don't need a PhD in Meteorology to know what the author is talking about, and that is important when the subject is as potentially life changing as climate change could be."

Kenneth Campbell, MD Big Green Book

"I regret it is just too painful a process to get my comments past our legal department!"

Name withheld

To Mary,
for her unfailing support throughout this project.

Contents

Will Climate Change Your Life?

Introduction

WILL CLIMATE CHANGE YOUR LIFE?

'We are past the point of no return'

All right, I admit it. You can't save the planet by driving a 4x4. But you won't save the planet by driving them off the roads either.

If global warming, climate change and so on are as dangerous as some people claim, maybe we shouldn't be driving at all. I don't need to go into detail to explain why we can't just stop driving tomorrow. We need to get to work, do the school run, go to the supermarket, visit friends and so on and so on. Public transport is not the universal answer. Cycling is fine when the weather is dry and not freezing, but only for local journeys. Ditto walking. For most of us in the prosperous West the car is a fact of life; not a choice, but a necessity.

If we need to stop driving to save the planet, this will not just affect cars. Anything which uses fossil fuels — petrol, diesel or LPG - will have to be cut back. That's vans, lorries and trains. And don't forget aircraft. No cheap flights to the sun. Maybe no flying at all, except for the very rich.

But how realistic is all this? Not at all, at least not in the short or immediate term. In the UK in 2000 we saw the effect of the fuel protest. A few farmers and hauliers blocked the oil refineries and after only a few days life got so difficult that the government gave in to their demands and cut the fuel tax. We certainly wouldn't volunteer to go back to that situation and, if it was somehow imposed on us, society would collapse. We cannot live our lives without fossil fuels — oil — to power our vehicles. We cannot live our lives, cook our food, heat our homes, run our PCs, access our bank accounts or do almost anything else without electricity, and most electricity in the UK comes from fossil fuels.

So why would anyone seriously suggest that we should consider such radical changes to the way we live our lives? (And some of these are serious scientists, not eco-freaks or anarchists.) Two reasons: climate change and energy shortages.

Climate change? Environmentalists believe that dangerous climate change will affect our lives in the next hundred years. We'll see forest fires, floods and famines, mostly in the developing world, but possibly in Europe and America as well. If we go on burning fossil fuels, they say, it can only get worse.

Energy shortages? Well the oil is going to run out, and the gas too. And even if it doesn't, we in the UK are increasingly dependent on overseas supplies, so we could have short-term supply problems if something goes wrong down the pipeline.

Climate change, as I said, will have an effect over the next 100 years and beyond. The next fuel crisis could happen tomorrow. If it doesn't, it will certainly happen in the next 10 years and bring changes that will be final, permanent and for ever. After all, if the oil runs out, it has run out; there isn't any more where it came from.

We hear about carbon emissions all the time. Some people, like George Monbiot, say we should cut our carbon emissions by 90%. That means, more or less, that we need to cut our energy use by 90%. Use it more efficiently, or cut out 90% of our journeys and 90% of all the electricity we use.

Most people ignore the messages of doom we read in the papers. After all, the Millennium Bug didn't bring life to a standstill in 2000; we haven't all died of SARS; and some 300 million Europeans have been using the Euro without many problems for the last five years. Why shouldn't we keep driving the 4x4, taking long-haul flights for foreign holidays and leave the TV, video, microwave and PC on standby?

There are also conflicting opinions:

US Senator James Inhofe, leader of American opposition to carbon controls, said "Climate change is the biggest hoax ever perpetrated on the American people."

Sir David King, British government chief scientist, said "Climate change is the most severe problem we are facing today".

It was James Lovelock, eminent earth scientist, who said "We are past the point of no return."[1]

If the scientists and politicians can't make up their minds, why should we worry? Most of us have enough to do with earning a living, bringing up children and keeping the credit cards under control.

And yet – maybe, just maybe, this could be the big one. After all, if you believe James Lovelock, climate change will wipe out the whole human race within the next couple of centuries, and life will get increasingly unpleasant in the very near future. Forget the grandchildren – most of us under 60 will have a ringside seat. So in case he's right, should we be doing something – perhaps just a bit – about it?

The aim of this book is to sort out the hype from the horror stories and to answer the questions which are increasingly being asked:

◊ Is there a crisis?

◊ Is it our fault?

◊ What can we do about it?

◊ What happens if we don't do anything about it – or even if we do?

I'm not a scientist, but I expect scientists to do the science and then explain it to the rest of us.

. . . .

[1] The Revenge of Gaia - Penguin 2006.

"You wouldn't understand" is not good enough, especially if we are talking about nuclear power, stem cell research or global warming. If preventing global warming means paying more for petrol, using the car less and no longer taking cheap flights, we need to be sure that such sacrifices are justified and will be effective. We need to understand the science so we can make up our own minds.

This book aims to be a rational, readable review for busy people. I've asked friends and colleagues to read the drafts, and I've gone back to alter and amend in light of their valuable insights. This means that you should find this straightforward, practical and 'jargon-free', as far as possible. If you don't, please write and tell me at...

mail@willclimatechangeyourlife.com!

The Way We Live Now

WILL CLIMATE CHANGE YOUR LIFE?

WILL CLIMATE CHANGE YOUR LIFE?

What can you do about saving the planet? And more important, what should you do? And should you feel guilty about what you do or don't do?

The climate issue is only part of a much wider environmental picture. The whole thing is immensely complicated and for politicians to pick on high-profile easy targets like 4x4 drivers is cynical, divisive and totally ineffective.

It's probably not a good idea to drive a 4x4 because oil prices are rising and it's going to get expensive. All right, I know that, if you can afford the big car, the cost of fuel is not important. But if there is a disruption to oil supplies it will drive the price up even further, which will cause shortages, at least in the short term. Miles per gallon will become more important than what the petrol actually costs! Taking 4x4s off the road will actually have a very limited effect. We will only make a difference if we all change our driving patterns.

There are a lot of things you can do to save energy and some of them - a few of them – may save you money as well. You'll find details in Chapter six. Actually, most of these things will cost you a great deal of money. However, most will reduce the CO_2 that your lifestyle has been putting into the atmosphere.

THE REAL ISSUE

The people at the IPCC, the Intergovernmental Panel on Climate Change, who brought you the Kyoto Protocol[2], have predicted that over the next hundred years we can look forward to:

◊ More violent weather;

◊ Droughts in some places;

◊ Floods in other places as sea levels rise;

◊ Reduced agricultural production;

◊ Millions of refugees trying to escape all these events.

Even the sceptics accept that the climate is changing and most people agree that the greenhouse gases, caused by human activity and industrialisation, are a significant part of the cause.

It is difficult to see what we as individuals can do about this situation. It's difficult to see exactly how it would affect us. We are all concerned about droughts and famines in far off countries and we do what we can to help. Our first concern, though, is always ourselves. And unless we can see something that is actually affecting us directly we do nothing, because there's nothing we can do.

Recently, a newspaper asked its readers what should be done about climate change. That week they printed twelve pages of letters. The thing about all of them is that they said that something should be done - preferably by somebody else! Ban this, tax that, stop people doing something else! The eco-extremists were out in force. I'm not sure that I would like to live in the authoritarian state that these people describe!

. . . .

[2]***The Kyoto Protocol to the United Nations Framework Convention on Climate Change*** *is an amendment to the international treaty on climate change, assigning mandatory targets for the reduction of greenhouse gas emissions to signatory nations. It was signed by 166 nations, not including the USA or Australia.*

The people at the IPCC have prepared 40 scenarios to demonstrate the things that could go wrong in the next 100 years. That's 40 different possibilities. Nobody knows exactly what is going to happen, but the consensus is that whatever happens will be unpleasant, dangerous and will seriously affect millions of people.

On the other hand, nobody can say for certain that reducing greenhouse gas emissions will slow down or stop global warming or have any effect on it at all. We are in a difficult situation. We can either put a tremendous amount of effort into cutting greenhouse gases and find in a few years time that it has cost us a great deal and achieved nothing at all, or we can save our energy and see how things are going to turn out and then find it's all too late to do anything.

If you read the reports from the 2005 Hadley Conference[3] you will see that scientists predict that if we stabilize the level of carbon dioxide now it may have a positive effect on the climate in about 100 years. During that time we will still see the storms, the droughts, the floods and the rising sea levels. Of course, that is not an argument for doing nothing at all. It's probably a good plan to cut carbon dioxide emissions to try and stop things getting any worse.

We need to realize that whatever we do, we need to protect ourselves in a violently and fundamentally changing world. Having said that, our lives are going to be affected much more immediately by the coming energy crisis.

Oil is running out. Yes, I know that's a scare story that has been around for years, but oil is a finite resource, it must run out sometime. It probably won't run out tomorrow or next week, or possibly for another 40 or 50 years. However, those who know, principally the Association for the Study of Peak Oil and Gas (ASPO) - a body of ex-oilmen - believe that we have reached peak production levels and supplies can only decline from here on.

....

[3] *"Avoiding Dangerous Climate Change"* Cambridge University Press 2006 *ISBN 0-521-86471-2*

At the same time, the developing nations, particularly India and China, are demanding more and more energy, so the price can only go up. Neither China nor India has significant domestic reserves of oil, so they have to bid against us and the other western nations for oil on the world market. For most of the twentieth century, oil has been traded at $20.00 to $30.00 a barrel. At the time of writing (mid 2006) the price has been up to nearly $80.00 a barrel. It has fallen back to around $60, still 100% above the long-term trend.

Britain is in a better position than a lot of countries, but arguably much more at risk than it was 25 years ago. In 1980 we were self sufficient in energy. In 2006 North Sea oil is running out and we are starting to import oil again. North Sea gas is running out. We import 10% of our gas requirements and the forecast is that this will rise to 80% within a decade. Our coal industry is smaller than it's ever been: some 60% of our coal is imported from as far away as Russia, South Africa and even Vietnam! Britain's energy security is no longer under Britain's control. When the lights go out we all go out of business, and it won't be much fun sitting at home in the dark!

How are we going to meet all these challenges? First we need to define exactly what they are.

The Green Red Herring

There is a strong possibility that climate change will have the greatest effect ever seen on human civilisation. So why is this a red herring? Because there is a problem that is more immediate, more dangerous and more certain.

Climate change may be devastating for the developing world over the next hundred years:

The energy problem will devastate the West in the next decade.

Cutting greenhouse gas emissions and the production of carbon dioxide (CO_2) is a red herring because nobody who really knows – the meteorologist or climate scientist - claims that it will do any good for hundreds of years. Yes, CO_2 is a greenhouse gas and causes global warming. It has been produced naturally since plants have been on the earth. It is produced as plants and organic matter decay, as rocks are eroded by the weather and as we, and all other animals, breathe out. Since the start of the Industrial Revolution 200 years ago humans have been pouring more and more CO_2 into the atmosphere as a by-product from our factories, our power stations, our aircraft and our cars.

The UK currently emits 150,000,000 tonnes (or 150Mtc)[4] each year. The government's target – which it seems unlikely to meet - is to reduce this by 20,000,000 tonnes by 2012. Total UK CO_2 emissions are 2% of world emissions, and every tonne of carbon dioxide that we emit persists in the atmosphere for 100 years.

So should we do nothing? NO! The best way to reduce CO_2 is to reduce energy use, and reducing energy use is the best way to prepare for the coming energy shortages.

We should also prepare for the consequences of global warming by preparing for floods, forest fires, drought and famine.

We must be ready to help the developing nations, which are likely to be worse affected than the industrialised nations.

Above all, we must monitor the situation and adapt our plans as the situation develops.

Nobody knows how climate change will develop. The United Nations has presented a range of different scenarios for the next 100 years, because the actual outcome will depend on how governments and industries react, how the oceans adapt, how the ice-caps behave and how vegetation affects the heat absorbed by the earth.

As we monitor and learn we will be better able to prepare…

. . . .

[4]Greenhouse gas emissions are expressed as million tonnes of carbon equivalent (MtC). One tonne of carbon is contained in 3.67 tonnes of carbon dioxide which is the ratio of the molecular weight of carbon dioxide to the atomic weight of carbon (i.e. 44/12). Other gases are expressed in terms of carbon equivalent by multiplying their emissions by their global warming potential (GWP) and dividing by 3.67.

Somebody stopped me here and asked, "What's all this 'we'? Who are you talking about?"

Let's make it quite clear; 'we' is you and me, on behalf of that endangered species, the human race. "But surely," some people say, "it's up to the government." It's true that these issues are so large that only action at government level can make a real difference, so we are going to have to persuade our politicians to take action. I know a lot of people see politics as a no-no, but politicians are acting in our name, with our money. It's what politics is for. This is not about party politics – all parties should be made to agree on something as serious as this. My use of 'we' is designed to make people realise that the future of the planet is down to us and absolutely no-one else. Let's continue.

…We have already learned that reducing CO_2 may reduce the effects of climate change in centuries to come. We also know that cutting CO_2 emissions now will not stop global warming or its immediate consequences. That is one of the most important (and controversial) statements in this book, so here it is again:

Cutting CO_2 now will NOT stop global warming or its immediate consequences

SO, IS THERE A CRISIS?

The risks of serious consequences from climate change and global warming are real. Chapter two goes into some detail on how global warming occurs, how we know it is occurring and what the likely consequences may be.

By far the most authoritative document is the Third Assessment Report (2001) issued by the United Nations' Intergovernmental Panel on Climate Change. It describes how global warming has occurred, but it also shows how the effects of that global warming cannot be accurately predicted. As this book goes to press the IPCC is publishing its Fourth Assessment Report. For commentary on this please go to *www.willclimatechangeyourlife.com*

There is tremendous uncertainty. In 2005 people claimed that Hurricane Katrina, which laid waste to New Orleans, was a symptom of unstoppable climate change. In 2006 there have been no hurricanes worthy of the name and some scientists have now made it clear that they doubt that the warmth of the ocean had anything to do with Katrina's strength.[5] Yet if you listen to the popular press or campaigners like Al Gore you will believe that we are in the grip of a crisis. A crisis that can only be solved by immediate and dramatic cuts. The Camp for Climate Action held in North Yorkshire attempted to achieve cuts by closing down Drax, the UK's biggest coal-fired power station.

Anyone who opposes this approach is vilified as irresponsible, complacent or, worse still, in the pocket of 'Big Business'. In fact, the most irresponsible and complacent are those who claim that all we have to do is cut CO_2 and the problem will be over. These people seem to ignore that in the UK we produce only 2% of global emissions, while the US produces 21% and China, at 15%, is well on its way to double its output.

True complacency is failing to demand a foreign policy that opposes the polluting nations, or at least encourages them to use clean technologies.

True complacency is doing nothing to help those poorer nations which will be the first to suffer from forest fires, floods and famine.

It's important to recognise that the consequences of global warming are real and will affect us for hundreds of years. It's also critical that we recognise that there is nothing we can do to stop it immediately, even though there are things to be done now which may limit its future effects. There are things to be done to help those most likely to suffer the consequences.

There are those who deny global warming, though their number is declining. There are also those who advise caution, and urge us to monitor the situation and tailor our responses to the way that climate change is seen to develop. This seems to be a perfectly reasonable

....
[5]*Chris Landsea, US National Oceanic and Atmospheric Administration*

approach, but these people, too, are attacked and shouted down by the eco-campaigners who believe that the only true way is CO_2 reduction. They clearly state that this will be a welcome step towards smashing capitalism. You'll find more about the views of some who reject global warming and some who reject our reactions to it in Chapter three. See who is saying what and then make up your own mind.

ENERGY SHORTAGE

Global warming is, of course, global, but different parts of the world will be affected in different ways. Climate change, the consequence of current events, will affect us in twenty, fifty, 100 years time. In the short term, however, the answer to the question 'what crisis?' is more likely to be our energy shortage.

Energy use is bound up with the whole climate change issue – burning fossil fuels releases CO_2 which makes the greenhouse effect worse. But in the short term, energy use is what defines a whole civilization and keeps it going.

If the lights go out we all go out of business. If we have a serious, prolonged power cut how many supermarkets do you know with windows that let in the daylight? Or freezers, or tills, that don't need electricity? Most supermarkets and shopping centres have a diesel generator tucked away at the back of the car park. Some even have wind turbines now. If there was a short term power cut you probably wouldn't notice. But after two or three days that diesel tank would run out and there might be no wind.

Our factories, offices, schools and hospitals all depend on energy. Energy pumps our water and drains away and treats our sewage. Energy powers our phones and runs the cash-points. It heats and lights our homes. It runs our transport system. Energy is our life. Relatively minor disruptions caused by the 2000 fuel protests caused serious problems for many people. An extended shortage would strain our civilisation.

Secure supply

How secure is Britain's energy supply? In 1980 we had a thriving (if filthy and dangerous) coal industry. It fed our power stations which produced 73% of our electricity. Another 14% came from nuclear. We had North Sea oil and gas and we were close to self-sufficient in energy.

Now look at the present situation. There's more about how our coal production has declined and imports have increased in Chapter four. We still use coal for power stations but 50% of coal for Drax (the UK's biggest) comes from Russia because it's cheaper.

In 1980 we imported 6% of coal: in 2004 that had risen to 59%. In the eighties and nineties the government decided that the public would not accept any more nuclear power stations so they introduced the 'Dash for Gas'. Gas fired power stations are quick and cheap to build. They don't pollute nearly as much as coal fired stations and they are flexible in use – they can be turned quickly on or off. Of course they need gas, and we have enjoyed the benefits of North Sea Gas. Enormous reserves lasting well beyond the life of the parliament that brought in the dash for gas. Not lasting for ever, though. In fact North Sea gas – and oil – are beginning to run out, but gas already accounts for 34% of electricity generation – and growing, while coal now produces only 38%.

We are importing 10% of our gas requirements and by 2015 that's expected to rise to 80%. Gas is imported from Norway and the Netherlands and can also be brought in tankers by sea from Qatar. Russia, with some 27% of the world's natural gas reserves is likely to become the principal supplier to the UK and the rest of Europe. At the end of the day, though, gas is traded on world markets and we have to compete and pay the price demanded. Whatever the rights and wrongs of the case, we saw in early 2006 how Russia could impose a gas price increase on Ukraine by cutting off supplies. And the UK is at the very end of a very long pipeline, so if Russia should have a dispute with any of the countries along the way, our gas supplies could rapidly be at risk.

Conventional sources of energy, how long they will last and how they can be supplemented are explored in Chapter five.

Chapter six assesses the viability of alternatives to oil, gas and coal for energy – including nuclear, renewables and hydrogen – and whether we can set these up in time

In the rest of the book you'll find explanations of the sort of things that people suggest we can do: how we can influence events and the sort of steps we can take to protect ourselves, our families and our businesses without going bankrupt or living on bread and water in a tent.

The next chapter examines the causes and consequences – as well as the certainties and speculation about global warming. Actually, it seems to me that we will experience problems from energy shortages much sooner than climate change will change our lives. But saving energy is one of the most effective ways of saving CO_2!

So don't let's ignore global warming, but let's get our priorities right!

CO$_2$ and climate change

Global warming supports life, but if it ever got out of balance life could be destroyed. There are signs that the climate is becoming unstable and signs that human activity is part of the cause. Fire, flood and famine may flow from all this, so what can we do?

Several ideas are proposed, though no-one knows exactly what's needed, or how much or how soon. Does the global community have the political will to act together to do the right thing?

THE CLIMATE SYSTEM

A thin layer of atmosphere covers the surface of the earth. The gases and the water within this layer are what make life possible. Together with the sun, they make the weather. Ocean currents carry heat around the globe. Clouds carry water from place to place, blown along by the wind. Clouds shield the earth from much of the sun's heat, preventing it from overheating. They insulate the earth like a greenhouse. They reflect heat back to the ground and prevent it escaping at night. They stop life on earth from freezing to death.

For millions of years and for reasons that are still not clear, the climate has remained in balance, temperatures have been broadly stable and life on earth has developed and prospered.

It's not just water vapour and clouds that hold in the heat. Other gases like methane, CFCs (chlorofluorocarbons), CO$_2$ (carbon dioxide) and SO_2 (sulphur dioxide) all play an important part. CO_2 is the most common of these.

Recently scientists have detected an increase in the average temperature of the earth; about 0.5°C in the last 100 years. At first sight half a degree is negligible, but they believe it is the first sustained change in temperature for 40,000 years.

Based on this rising trend researchers have created scenarios demonstrating an increase of up to 5.8°C by 2100. At the same time they have found sharply increased levels of CO_2 in the atmosphere; levels which started increasing at the beginning of the industrial revolution, when man began to release CO_2 by burning ever greater quantities of coal, oil and gas. The conclusion is that if we continue to burn these fossil fuels and continue to burn them in larger and larger quantities, CO_2 levels will rise; the planet will heat up and may eventually become uninhabitable.

Some say that we have already passed the point of no return for the human race; others believe there is still time to take drastic action to solve the problem. Global warming is not unique to the earth; it has also been observed on Mars and Venus. The atmosphere on Mars is very thin so the heating effect is small, but the surface temperature of Venus is more than 500°C. This is due to a very much thicker atmosphere on Venus than the one we have on earth and to the fact that Venus is hotter because it is much closer to the sun. Such excessive warming could not happen here, but scientists predict that even one or two degrees could have a major effect.

The last Ice Age finished some 20,000 years ago. In the 1970s following a series of cold winters, the press was predicting that the next Ice Age could start any minute. Now, after several hot summers in the 1990s and life-threatening heatwaves across Europe in 2003 and 2006, global warming is the popular fear.

Scientists are confident that global warming and climate change are real and that they are due, in part, to human activity. The difficulty is that, while they may accept that their projections are only projections – *possible* scenarios, the popular press see them as predictions or established fact. Many people accept them as such without question. Scientists are telling us what might happen, but too many people believe that it really will happen.

The climate is an intensely complicated system, changing and developing with the unpredictability of catastrophe theory. Undoubtedly it changes slowly, so two or three unusually hot or cold years are no indication of how that change is going.

Looking at evidence of the climate over thousands of years, scientists have concluded that the earth is warming more rapidly than at any time in the last 10,000 years. Taking the average of the IPCC scenarios, (the projections made by the scientists for the United Nations' Intergovernmental Panel on Climate Change), global average temperatures will rise by 3° over the present century, though not evenly over the whole surface of the globe. This compares with a rise of 0.6° since modern temperature records began in around 1860 and with the difference of 5° or 6° between the present climate and the coldest period of an Ice Age. This 0.6° increase may appear small, but it is seen as a sharp rise when compared with trends over periods as long as 400,000 years or more.

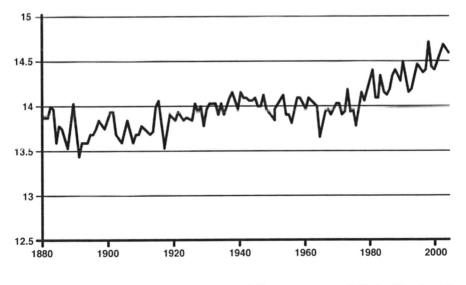

Average Annual Temperatures (°C) 1880 - 2004
Source: Goddard Institute for Space Studies

How can we know what the temperatures were so long ago?

◊ Tree rings indicate annual temperatures – they grow faster and thicker in warm weather. By drilling cores out of old timber and measuring their thickness it is possible to track the temperatures year by year. Relatively new timber can be matched with the temperature measurements from thermometer readings, and then, by comparing overlapping samples of timber, annual temperatures have been tracked back some 8,000 years.

◊ Ice cores have been drilled out of the Arctic, containing the layers of snow which have fallen over the last 400,000 years. Analysis of the gases, chemicals and pollen trapped in each year's layer indicates the annual temperature.

◊ Ocean sediments have built up slowly over time and analysis level by level takes the record back a million years.

These techniques provide a long-term temperature profile for the earth and a basis for deciding whether present climate change is unusual.

Charting the trend

The record of the Greenland ice-cores indicates that the climate has been unusually stable for the last 8,000 years. Since the last Ice Age the climate has warmed by 0.2° per century. For the 20th century the increase was 0.5° and for the 21st century a rise of 2.5° – more than ten times as much – is considered more than possible.

One of the most important things to remember is that we are not looking at a smooth and steady progression. The trend over the last 10,000 or 100,000 years is the line which gives the statistical 'best fit' to a range of results, up and down. Just as now, we have had bad winters and exceptional summers from time to time throughout history. The trend allows for these short-term fluctuations and demonstrates whether there is any significant long-term trend. For this reason a short-term fluctuation on its own tells us nothing about climate change.

The hottest July in England since records began is no proof that we are in the grip of global warming. Was it very hot in Europe? In the rest of the world? Remember, in May 2006 we were complaining about how cold it was and how late spring was in coming! There is evidence that in only 50 years the Arctic warmed by 7°C about 10,700 years ago and during the last 100,000 years there have been fluctuations in Greenland of as much as 16°C. Yet the long-term trend was only 0.2 °C per century as we have seen.

The changing climate

The effect of climate change depends on so many variable factors that it is impossible to *predict* what *definitely* will happen. It is certainly possible to *project* what *might possibly* happen, and this is why the IPCC Third Assessment Report contains some 40 different scenarios, each based on a different range of assumptions.

The four key storylines are

A1: Rapid economic growth, rapid peak in population growth followed by a decline; new and more efficient technologies

A2: Slower economic and technological development; constantly rising population

B1: Changes towards a service and information-based economy; a rapid peak in population growth followed by a decline; clean and resource-efficient technologies

B2: Slower economic growth; constantly rising population; less rapid and more diverse technological change

Each of these storylines leads to a family of scenarios, each of which produces a different outcome in terms of climate change on the basis of different assumptions.

It is clear that, although the panel has developed a good range of possible predictions for the future, there is no certainty at all of what the eventual outcome will be. It demonstrates that it is as important (and difficult) to predict the actions of governments and people as to predict the reactions of the climate system.

Climate change will have an impact, to a greater or lesser extent, on the global environment. There is no doubt that the effects will be unevenly spread around the world and the signs are that the less developed nations will get the worst of it. The IPCC scenarios predict sea level rise, averaging 0.4m by 2100, but possibly as much as 0.8m.

If the rise is 0.5m, this will be enough to cover 10% of Bangladesh and displace 6 million people. A rise of 1.0m would cover 20% of Bangladesh and displace 15 million people. Although the global average sea-level rise may be only 0.4m, local factors mean that Bangladesh could see a full 1.0m rise as early as 2050. It cannot absorb 15 million people on to the land that remains and it cannot replace the agricultural production of the lost land from its own resources. The only consequence will be mass migration of refugees.

While sea levels will rise steadily over the coming years, the increasing base levels mean that storm surges will be more serious and floods will extend further inland. People will not be displaced by gradually rising waters; they will be driven out by catastrophic events with extensive loss of life and property damage.

And Bangladesh is not alone: a one-metre rise is also likely to affect Egypt by 2050, flooding 12% of the land and displacing 7 million people. China, the Netherlands, the Southern US states (remember New Orleans?) and the Maldives among others will all be affected. Once the storms have broken through there is a good chance that the waters will never flow back. Even if they do, the salt left behind will ruin the land for agriculture.

Rising sea-levels reduce access to groundwater – salt water pollution will be found at shallower and shallower levels. In many countries there is already severe pressure on water supplies. Over-extraction of groundwater causes subsidence and makes land more at risk from flooding.

Even if global temperatures are stabilised, sea levels will continue to rise for many years to come. Much expansion is due to the fact that warmer water takes up more space. The oceans are so huge that they heat up and cool down very slowly. It can take thousands of years for changes in the surface temperature to affect the whole body of water deep down. Even if we could halt global warming immediately, climatologists believe that sea levels will continue to rise for at least 1,000 years.

Rising global temperatures will lead to increased rainfall, but only in some parts of the world. Climate models indicate that more rain is likely in the far north, with smaller increases for areas including northern Europe and North America. Further south, in southern Europe and North Africa, rainfall will be lower and temperatures significantly higher. Higher temperatures and lower rainfall will also affect areas such as Central America and Southern Africa. Elsewhere changes will be smaller or more difficult to predict.[6] The higher temperatures will cause more evaporation so more of the rain that does fall will be lost.

With a shortage of fresh water there will be lower agricultural production, more droughts and famines, more refugees forced away from their lands. Political tensions will rise where countries depend on the same rivers for supplies. The Danube passes through twelve separate countries, the Nile through nine. Already there are disputes between Egypt and Ethiopia over the use of the Nile waters. Ethiopia is a poor country with little rainfall. It wants to divert water from the river to irrigate fields and boost agricultural production. Egypt opposes this, as any water used by Ethiopia will reduce the waters flowing down to Egypt where 85% of the river's capacity is already used for irrigation. If rainfall decreases and river levels fall, disputes over water can only get worse.

The Aral Sea is a graphic example of a water-related disaster. If you have seen Al Gore's film you will remember the pictures of the fishing boats abandoned in the desert. Once the fourth largest body of inland water in the world, the sea lost three quarters of its water

....

[6]*Global Warming – the complete briefing: Houghton*

by volume and half its surface area between 1960 and 1995. The remaining water became so salty that fish could not live and the fishing industry collapsed. The whole ecosystem of the area was destroyed and the lives of 5 million people affected by the dust of the desert, the contamination of groundwater, the failure of agriculture and the spread of disease.

The collapse of the Aral Sea was not a result of climate change; it was wholly man-made. The two rivers that fed the sea were diverted to provide irrigation for cotton crops, and without them the sea just dried up. If climate change causes rains to fail and river levels to fall, the Aral Sea provides a stark example of the consequences for the world.

To some extent, farmers can adjust to changing weather patterns by adjusting the time when they sow and harvest and by changing the type of crops they grow. Farmers in the developed world will have more skills and resources to experiment with different planting patterns and crops.

All plants absorb CO_2 and release oxygen. Some plants are stimulated by additional CO_2. Increased CO_2 can produce increased growth, but poorer quality may result. The evidence currently available indicates that global agricultural production will remain much the same, but the distribution will change. While some areas will profit there will be other areas of extreme poverty.

If climate changes at the rate foreseen – rising by an average of 2.5°C in this present century – there will be a profound effect on natural ecosystems. This means that many species of trees and plants will die off when their normal environment becomes too hot or too wet for them. Animals and birds that are unable to adapt will become extinct in some regions. Ecosystems can adapt, but not to the rapid rate of change that some expect. For example, cod are already found further and further north where the water is colder, but trees and plants cannot walk or swim so can migrate only as new generations. Plants at the south of an area will die off and more plants will grow at the northern edge. If the whole area becomes too hot for a species before it matures then it will all die back. Some forests are already affected like this by climate change – and by pollution. Whereas the great forests of the world absorb thousands of tonnes of CO_2, dying forests release CO_2 to the atmosphere. Where forests die, the animals and birds that live there die out as well.

Some ocean fish can adapt by migrating to different waters, but this will have implications for fishing and for communities who rely on fish for their diet. The effect of 'El Niño' is already well known. This is a warm current flowing up the west coast of South America. As long as it flows well away from the coast, fish are plentiful. From time to time it changes course, flowing close into the coast and the fishery collapses for years at a time. The fish are swimming deeper or further out to sea. Different species of fish are affected in similar ways right up as far as the California coast. Climate change will bring other shifts like this, driving the fish too far away for fishermen to catch them.

Human health

In 2003 there was a heatwave across Europe; this was particularly acute in France. With daily temperatures exceeding 40°C, the death toll for August was nearly 15,000 more than normally expected for that month. Deaths were mainly among elderly people who succumbed to overheating and dehydration. Humans are adaptable and, if such hot summers become the norm, people will become more tolerant of them and will be better able to prepare for them.

Warmer winters will have the opposite effect on mortality: With fewer frosts and cold winds more people will survive. While it may be possible to adapt to the effects of climate alone, if climate change has damaged agriculture and people are short of food and drinking water they will be more vulnerable to extreme temperatures.

They may be weakened by disease as climate change widens the range of disease-carrying insects. There is argument about how significant the disease threat is; since malaria was found in northern latitudes before the climate began to warm.

As with heatstroke, the vulnerability of a population depends on its general state of health, nutrition and hygiene. This is another indication that those who will suffer most will be the people of the developing countries. Actions to reduce these consequences include:

◊ Analysis of the risks and vulnerability of water supplies

◊ Improved training and emergency management

◊ Early warning systems for floods and other disasters

◊ Development of victim support and stress counselling

All these things have to be put in place. While actions to stop climate change may work in the long term, we need to handle the short-term consequences that we cannot prevent.

Other events

There are some events which are so enormous that on their own they would have a catastrophic effect on the whole world and the global population.

A shut-down or a 'flip' of the THC could bring Arctic winters to the British Isles and much of Europe. The THC is the *thermohaline circulation,* or what is loosely called the Gulf Stream. In fact the Gulf Stream is only part of a system of currents circulating around the world. The Gulf Stream is heated in the Caribbean and flows north past Britain and Ireland until it reaches the Arctic. At this point it cools, becomes denser, sinks to the sea bed and flows back to the Caribbean. The warm air it brings as it travels north prevents cold air from spilling down over the UK. This is why Britain has winters which are generally mild and wet, while winter in Labrador, which is no further north, is frosty, icy, snowy and cold.

If the THC were to stop or to flip – start flowing in the opposite direction – Britain could freeze. It is known that the current did flow in the opposite direction thousands of years ago. It is also suggested that as the Arctic icepack melts, the fresh water released will dilute the Gulf Stream as it arrives making it less dense and slower to sink to the ocean floor. This will reduce the effect of the whole circulation system and the warm air that it brings from the south.

Whether this will happen is impossible to predict. If it did, it could happen relatively quickly and there is nothing we could do about it. The balance of opinion is that it won't happen, so there is no point worrying about it.

Another catastrophe of global proportions would be the break-up of the West Antarctic Ice Sheet (WAIS). There are signs that this is beginning to happen, as the amount of melt exceeds the snowfall for three glaciers. The deficit is about 60%. If the whole WAIS melted, it would raise global sea levels by 6m, though it was initially considered that this could take 1,000 years or more to happen. There are now concerns that this could occur much more quickly.[7]

Even without changes of catastrophic proportions, predicted effects are expected to displace three million people each year. That means there will be 150 million refugees by 2050.

Certainty or speculation?

Any predictions about climate change are complicated by feedback and regional variations and also the reliability of historic data. The climate models have limitations –the behaviour of clouds, the ocean circulation, rainfall and the polar ice sheets are all speculative. Given that there are basic uncertainties in the calculation of the effects there must be uncertainties about the impact of climate change. Nevertheless, impacts are certainly expected and, as we have seen they're likely to be in the areas of water supply, food production and distribution, and changing sea-levels.

Climate models provide guidance while there is still much uncertainty. Research continues. In particular the carbon cycle of

....

[7]'The Antarctic Ice Sheet and Sea Level Rise' – Chris Rapley, British Antarctic Survey, published in 'Avoiding Dangerous Climate Change', Cambridge University Press.

the marine biosphere is not fully understood. The effect of climate change on people needs to be studied. Equally the effect of people and their reactions to climate change will, themselves, have an impact on the outcome of climate change.

Any action we take to correct or modify our environment has very long term consequences. One of the major difficulties is that politicians always have short term objectives. At least they are beginning to pay lip service to sustainable development. In summary this amounts to meeting the needs of the present without compromising the ability of future generations to meet their own needs. A number of criteria have been set up as a framework for movements towards sustainable development.

◊ To act on the best science

◊ To observe the precautionary principle

◊ To beware of non-renewable or irreversible consequences

◊ The polluter pays

◊ There must be social justice between countries and between generations.

Costing change

The costs of climate change fall into three categories.

1. First, there is the damage directly caused by the impact of climate change, such as the floods from sea level rise, the forest fires sparked by higher temperatures or hurricane and storm damage.

2. The second cost is the cost of adaptation: taking measures in advance to protect against the consequences of climate change. Examples are flood defences or the establishment and training of rescue teams.

3. The third cost is the cost of mitigation. This is the cost of measures taken to try and stop climate change happening at all.

Adaptation is generally more expensive for developing countries, so, yet again, we see the areas most at risk from climate change being least able to protect themselves.

Furthermore, as carbon dioxide concentration grows in the atmosphere the violence of the impact of climate change and the damage caused is expected to be disproportionately greater. A fourfold increase in carbon dioxide concentrations may lead to eight times as much damage.

It is extremely difficult to put a monetary cost on the impact of climate change. Assessing the cost is complicated by traditional methods of financial analysis based on interest rates – the cost of money. Is any interest rate relevant when the consequences may include the total loss of certain countries, overwhelmed by rising sea levels? If interest rates are not used, how do we choose between one way of tackling the impact and another?

To put an actual money cost on the consequences of climate change and the extreme events that are likely to arise is dependent on provisos, assumptions and caveats. Sir John Houghton[8] has, nevertheless, calculated that it could be of the order of 1.5% - 2% of global GDP (Gross Domestic Product) per annum. That's 2¢ in every dollar or every euro: 2p in every £1, just to clean up disasters. Bjorn Lomborg, whose views are examined in Chapter four, estimated the cost at twice as much: 4% of GDP. The figures in the 2006 Stern Review[9] are back to 1%, but the projected costs of doing nothing are enormous.

Finding the solution

If the risks and the consequences of climate change are as bad as some suggest, the remedies could be devastating. Most of this cost

....

[8]*Global Warming - the Complete Briefing: third edition*
Cambridge University Press 2004
[9]*See Chapter 7.*

will be incurred in the developing world; in countries too poor to pay. Chapter seven presents some ideas for what could be done.

Before we get into that though, not everyone agrees that climate change is a problem. Others say we should spend more time on dealing with the consequences than trying to stop something that we still don't understand. See what they say in the next chapter.

Battle Lines

WILL CLIMATE CHANGE YOUR LIFE?

All those against!

Sir David King, Chief Scientific Advisor to the British Government, says "Climate change is the most severe problem we are facing today." James Lovelock, scientist and author of the Gaia theory of the earth as an organism, says "We are past the point of no return".

◊ On the other hand, US Senator James Inhofe has called climate change "the biggest hoax ever perpetrated on the American people." So who is right?

◊ The United Nations IPCC (Intergovernmental Panel on Climate Change), brings together scientific opinion from all over the world. We have NASA, Oakridge Laboratory, the Hadley Centre, the National Academy of Sciences, DEFRA, the Department of Trade, the European Union, all firmly convinced of the dangers of climate change.

◊ There are articles in Nature, New Scientist, newspapers and hosts of TV programmes.

◊ Then there are campaign groups, like WWF, Greenpeace, Friends of the Earth, and the Green Party.

◊ Al Gore, former US presidential candidate, has toured the world with a slide show which he has now made into a film and a book.

There is no doubt that climate change and global warming have developed a very high profile in recent months. The danger is that the issues have become oversimplified. Whatever the public or popular press may believe, the actual consequences of global warming are far from clear.

The IPCC's 40 different scenarios are just the tip of the iceberg. Research on refining these predictions continues all over the world and the IPCC will publish its *Fourth Assessment Report on Climate Change* in 2007.

The future is not clear – it never is – but it would be just as reckless to assume that there's a simple solution as it would be to do nothing at all. Many people believe that all we have to do is to cut carbon dioxide emissions and the global warming problem will be solved for the whole world. Anyone who suggests we should question the science or try something different is loudly shouted down, at least in Europe. In the U.S. the body of opinion that believes that global warming is a hoax, untrue or irrelevant is still in control, despite the protests of Al Gore, Arnold Schwarzenegger and others. As we shall see, the U.S. Senate was persuaded to reject the Kyoto Protocol ostensibly on the basis of 'science'. However, that 'science' can be shown to have little foundation.

Making a case

In America, lobbying, public relations and opinion-forming are more intense than anywhere in the world. PR experts and campaign managers are wheeled out at election time: the rest of the time they work for major corporations and organizations. Their objective is to influence public opinion and to influence the government to pass legislation favourable to their clients or to water down restrictive laws.

As the journalist said, 'never let the facts get in the way of a good story'. In many cases the facts are generally agreed, but lobbyists exploit the slightest uncertainty to raise doubts in the public mind. The clearest example of this has been the battle by the tobacco industry to escape responsibility for smoking-related illnesses. For decades they fostered doubts about the evidence and blocked attempts to put warnings on packets. They sponsored organisations with important sounding names that issued press releases, gave interviews and made announcements. More recently they published reports on the internet and encouraged others to copy them so that their reports would always outnumber any others picked up by an internet search.

When an author submits a paper to an academic journal such as *Nature* or the journal of one of the academic institutes, the editor selects a panel of experts in the field and asks for their comments on it. They look at the methodology, check the calculations, investigate the sources and compare the author's work with their own expert knowledge. They may go back to the author and ask for more evidence or clarification. If they are satisfied that the content and conclusions of the paper are original and justified then the editor will publish. This process of peer review ensures that only articles based on sound science and up to date knowledge are ever used.

Organisations set up with the help of lobbyists do not use peer review, but publish as widely and loudly as possible. The objective is to make sure that most of the information generally available, quoted and re-quoted, supports their view.

Repetition can never make anything true, but it can make people believe it. For example, Alexander Fleming did not discover penicillin as a treatment for infectious diseases, Thomas Edison did not invent the light bulb and James Watt did not invent the steam engine, despite what most people believe.[10]

Points of view

Here are the arguments of four people who do not accept the popular view of climate change, an author, a politician, a statistician and

....

[10]*Fleming discovered penicillin, but Howard Florey and Ernest Chain, who shared the Nobel Prize, discovered that it was a treatment for disease. After innumerable trials, Edison licensed a successful design for a light bulb from Joseph Swann of Edinburgh. Thomas Newcomen was installing steam engines for draining Cornish mines years before Watt built his engine, (though arguably these were atmospheric rather than true steam engines). It all shows the power of publicity on popular belief.*

an academic. Some are easily dismissed; others deserve to be taken seriously. The author and the politician deny that global warming is a problem – and the politician seriously doubts that it even exists. The statistician and the academic are more concerned with the way we have interpreted the data and the steps we should be taking.

Let's summarise their views.

◊ The author is sceptical: *global warming is a fad.*

◊ The politician is dogmatic: *global warming is a hoax.*

◊ The statistician is pragmatic: *the data could be interpreted in different ways, so let's not invest in solutions until we are truly sure they will solve the problem.*

◊ The academic is prescriptive: *global action is prestigious but unlikely to be effective. We should face up to unwelcome reality and protect ourselves at the local level against things we cannot change.*

THE AUTHOR

Michael Crichton? Isn't he a novelist? Yes, but he has apparently taken great pains to place his novel *State of Fear*[11] on a solid scientific basis with a detailed bibliography running to 20 pages. It is, nevertheless, important to check the credentials of these sources. Critics also claim that Crichton has been selective in only picking out evidence which supports his view. It is said that the book has strongly influenced the opinion of George W Bush on the environment.

The book takes the position that pollution is a much greater threat to the world than climate change. Crichton also doubts the science or, indeed, doubts that it is proven. He questions that the conclusions that people draw from it *can* be proven. He is very concerned at the politicization of science. He belongs to the group that are concerned

....
[11]State of Fear *Harper Collins 2004 ISBN 0-00-718159-0*

that climate disaster is a fashionable subject and, therefore, attractive to academics because grants in the field are readily available.

He draws a parallel with eugenics in the early part of the twentieth century. For 50 years celebrities, politicians and intellectuals believed that eugenics, or selective breeding of humans, would be essential to the survival of the human race. It was only when Hitler and the Nazis put the theories into practice that people realized how misguided and abhorrent they really were.

Crichton claims that the issue of climate change is the same as eugenics, insofar as the whole area is infected with fashion, hysteria and unreasonable expectations.

THE POLITICIAN

Senator James Inhofe was chairman of the Senate Committee on Environment and led the opposition to the ratification of the Kyoto Treaty by the United States. The Kyoto Treaty resulted from the second report of the Intergovernmental Panel on Climate Change (IPCC) and set targets for the reduction of carbon emissions by all the developed economies. Developing economies, such as China and India, were excluded from restrictions on the ground that they were very far behind the developed economies and it would be unfair to restrict their efforts to improve the standard of living of their populations. In a long and detailed speech[12] to the Senate, Inhofe announced that his case was based on three principles:

◊ objective science,

◊ consideration of the costs for business and consumers, and

◊ recognition that bureaucracy was the servant, not the ruler, of the people.

As far as objective science was concerned, he claimed that there was wide-ranging disagreement within the scientific community on

....
[12]For the full text see: **http://inhofe.senate.gov/floorspeeches.htm**

whether human activity was responsible for global warming and over whether those activities would precipitate natural disasters. For him, the pro-climate-change lobby were no more than 'alarmists.'

He cited some impressive names.

Dr Steven Schneider,[13] who is by no means a climate change sceptic, said that the IPCC scenarios should be considered in terms of the likelihood of their occurrence. He believed that there was a less than a 1% chance of the temperature increase being the maximum 5.8° and a 17% chance of it being less than 1.4°. Presumably that meant there was an 82% chance of the actual temperature increase lying somewhere in this range, a significant increase, so it is actually difficult to see how this supported Inhofe's case.

Inhofe then quoted Dr James Hansen of NASA who stated, he said, that 30 times the CO_2 emission reduction foreseen by the Kyoto agreement would be needed to reduce global warming to an acceptable level. Inhofe pointed out that if the proposed Kyoto levels would damage the U.S. economy: the effect of the 30 times that level would be unthinkable. However, Dr Hansen has since expressed dissatisfaction with the way his views were presented to the Senate Committee.[14]

Dr Frederick Seitz is a past president of the National Academy of Sciences, and Inhofe quoted the *Oregon Petition* issued by Seitz. This document claimed that CO_2 was actually beneficial to the environment and that the restrictions proposed by the Kyoto Protocol would be disastrous for the US economy. It was circulated to thousands of scientists of all disciplines, inviting them to sign it in support. The document was produced in a format and typeface which might have led the reader to believe that it was a publication of the National Academy. In fact, it was produced by someone with no appropriate qualifications; it was not peer-reviewed and was put together on a home PC. The National Academy of Sciences issued a warning and disclaimer stating that it had no connection with

. . . .

[13]*http://stephenschneider.stanford.edu/*
[14]*http://www.giss.nasa.gov/edu/gwdebate/ Hansen is no sceptic and complains that his views were inaccurately presented in evidence to the Senate Committee.*

the *Oregon Petition* in did not endorse it in any way. Hardly a firm foundation for Inhofe's case!

The *Heidelberg Appeal* was next put forward in support of the case against climate change. This document was widely supported by some 4,000 respected academics throughout the world, including 72 Nobel laureates. Some claim the whole thing was corrupt and fabricated by the same people who manipulated medical research for the tobacco companies. The truth, however, appears to be that the *Heidelberg Appeal* is a document calling for responsible objectivity in science, and does not specifically mention climate issues at all. However, climate change sceptics somehow managed to make people believe that it supported their position.

From the business point of view, the implication of the Kyoto Protocol was that the United States would have to reduce its greenhouse gas emissions by 31% of its expected 2010 level. This meant that the U.S. would have to eliminate all emissions either from all transportation or from all utilities or from all industry. Any change of this magnitude would clearly have an impact on jobs and on GDP; it would be unworkable. Who said so? The Western Business Roundtable.[15] The membership of this organisation is predominantly involved in energy, mining and coal. Given that these industries would be hardest hit by any restrictions on the use of fossil fuels, their view is hardly surprising. Their website shows that they continue to be sceptical of global warming and the efforts of the Europeans to control it.

The Europeans, after all, were heavily involved in all this. Comments by foreign leaders, said Inhofe, indicated that the real agenda behind Kyoto was global governments and foreign powers dictating U.S. government policy. Kyoto, he said, was really designed to restrain U.S. interests and to undermine the global competitiveness and economic superiority of the United States. "And this, because it will have no environmental benefits, while causing serious damage to the economy."

....

[15]**http://www.westernroundtable.com/** *The Western Business Roundtable membership is predominantly involved in energy, mining and coal. This link shows that they continue to be sceptical of global warming and the efforts of the Europeans to control it.*

So apart from economic sabotage, was there, in Inhofe's view, any other motivation for this enthusiasm for controlling the climate if there was no firm foundation in science? For him it was clearly the environmental groups seeking money and power. He had no doubt that there was plenty of money for research grants into the effects of climate change in global warming, and little money for those who took an opposite view. He claimed there were also extremists in the debate who simply did not like capitalism, free markets or freedoms and were setting out to destroy them all.

For US Senator James Inhofe then, the Kyoto Protocol was simply a conspiracy of foreign powers and others to handicap the American economy. This could not be allowed to happen!

The US Senate refused to ratify Kyoto. The case had been skilfully presented and the cause was won.

Some would call it a triumph of spin over substance.

THE SKEPTICAL ENVIRONMENTALIST

One of the most widely cited sceptics as far as climate change is concerned is Bjorn Lomborg, a Danish professor of statistics, who styles himself *The Skeptical Environmentalist* and has published a book with that name. Here is a person who may take a delight in the contrary view – he has strong opinions on many things beyond the climate – but someone who has no hidden agenda and represents no commercial or political lobby.

Lomborg has dared to challenge the popular wisdom and has been pilloried, by the *Scientific American* magazine among others. Lomborg is important because he is a scientist and has chosen to apply scientific method to the evidence, but has come up with an opinion which differs from the views of many others.

If the consequences of climate change are as serious as many expect, there is no doubt that the solution will be serious too. If we are trying to change things at the global level it is essential that the solutions we choose are the right solutions. Whatever we do at a global level could have unexpected side effects at a global level. Whatever we do has global risks. We only have one planet. We need to be certain that what we do to our planet is less risky than doing nothing at all. That is why it is so important to consider the contributions to the debate from clear and careful thinkers such as Bjorn Lomborg and, later in this chapter, Sonja Boehmer-Christiansen.

Defining the problem

As a statistician Lomborg challenges both the models and the methods of the IPCC analysis. He doubts whether we will see the extremes of some of the IPCC scenarios. He accepts the reality of man-made global warming, but does not believe that the maximum 5.8° temperature rise predicted by the IPCC is plausible. His position is that whatever we do is going to cost a great deal of money, and that we will achieve much more by tackling poverty and the other symptoms of global warming than by trying to reduce CO_2 emissions.

Lomborg sets out six questions which are fundamental to the understanding of climate change and to the choice of appropriate action:

1. How much effect does CO_2 have on temperature?

2. Could there be other causes of increasing temperature?

3. Are the greenhouse scenarios reasonable?

4. What are the consequences of possible temperature increase?

5. What are the costs of curbing versus not curbing CO_2?

6. How should we choose what to do?

It is fair to say that there are no simple answers to these questions and climatologists are constantly working to resolve them.

Lomborg casts doubt on the computer models used by the IPCC because of their scale and complexity. He questions whether all factors, including solar activity and sulphur dioxide in the atmosphere, have been given sufficient weight in the calculations. He questions the assumptions on cheap solar energy and government taxes on fossil fuels. However, his position is not to deny the consequences of climate change: his disagreement is a question of degree. There is a problem and we must take action, but the action taken must be the right action, and must be worthwhile.

Costing the consequences

We saw an assessment of the cost of climate change in the previous chapter and the latest view, the report to the Treasury and Cabinet Office by Sir Nicholas Stern[16] published in October 2006, is examined in chapter 7. Although by now Lomborg's figures may be out of date, his approach to the economics of climate change is still important.

Lomborg recommends that any action we take in attempting to mitigate the effect of CO_2 must be based on a trade-off between present costs and future benefits. In other words, the people (taxpayers) who have to pay to curb CO_2 must believe that the benefits created in the future will be worth it. He cites the Dynamic Integrated Climate Economy model and the Regional Integrated Climate Economy model (DICE & RICE). These models were developed by William D Nordhaus, Sterling Professor of Economics at Yale University in order to assess and quantify the costs of climate change.

On the basis of these analyses, Lomborg considers that 4% of Gross Domestic Product (GDP) would be the optimal level. This means that if we spend 4% of our GDP on reducing carbon emissions the losses we suffer now will be more than outweighed by gains in the future. However, if we spend more than 4% now, the extra benefits in the future will not be worth the extra sacrifice. It could even be argued that it would be easier and cheaper for us to do nothing at all now, but to wait for the future before making cuts. By then we may have better technology and be richer.

....

[16]*Stern Review on the Economics of Climate Change* ***http://www.hm-treasury.gov.uk***

This last point is a gamble, and most people prefer the precautionary principle: to do something now - in case we don't get richer and don't get new technology, and before it's too late.

Many people react violently against finance coming into the calculations when planning how to manage the environment. For them, the safety of our planet justifies action at any cost. Too often 'accountant' is used as a pejorative term, in the same breath as 'Big Business'. However, the truth is that everything in life is a choice, since resources are always limited. Expressing things in financial terms helps to weigh up one project against another. Finance and economics tells us what can be done; it is up to us to decide democratically what should be done.

In summary, the Skeptical Environmentalist has serious doubts that global warming is man-made, though he does not deny that global warming and climate change exist. He believes that we should only take steps to reduce carbon emissions if the results we achieve – now or in the future – justify the costs involved.

AN ACADEMIC PERSPECTIVE

Dr Sonja Boehmer-Christiansen is Reader at the Department of Geography, Faculty of Science, University of Hull[17] in the United Kingdom and Editor of *Energy & Environment*. During the course of a wide-ranging academic career she has studied environmental issues for some 30 years.

Her view is that global warming is no more than a bandwagon for many other agendas (ideological [green]), research-related (access to funding for 'policy relevant research'), commercial (subsidies, new markets), regulatory (harming competitors), political (enlarged competences, gaining allies and influence) which must be recognised and resisted as appropriate.

She does not dispute that global warming exists, or that man-made greenhouse gases have added to the effect, but doubts that we have a satisfactory scientific explanation for it and especially for future

....

[17] *http://www.multi-science.co.uk/ee.htm*

climate trends. She distrusts large global computer models, because they assume the man-made warming theory and hence cannot predict anything else.

She does not believe that reducing the use of fossil fuels will be possible without doing serious political and economic harm. The scientists themselves admit that the effects of emission reductions are unlikely to be experienced by anyone alive today.

So why the bandwagon? What is the motivation for dire, (she believes, overstated) consequences from global warming? If global warming has consequences she sees that there are two courses of action:

◊ Mitigation – taking steps to remove and reverse the causes of global warming,

◊ Adaptation – protecting ourselves against the effects of climate change.

Mitigation is the proposed global solution. But who proposes it? And why? Here a political interest analysis is needed.

The principal mitigation policy is the reduction of CO_2 and other greenhouse gases in the atmosphere of the whole world. This requires a degree of political and economic cooperation that is simply not forthcoming, as demonstrated by the long, but largely ineffective, negotiations under the Climate Change treaty and its Kyoto Protocol.

Adaptation is a local issue because the protection needed depends on the local/regional conditions. In some places there may be floods: better dams will be needed. In other places drought may require more irrigation channels or a complete revision of farming methods. Elsewhere the problem may be epidemics as warmer weather breeds insects and bugs. Adaptation is not a matter of global decision-making or regulation. There are no global solutions – though the rich may help the weak and there may even be compensation.

Dr Boehmer-Christiansen detects several political agendas as drivers, with science as justification or fig leaf rather than underpinning knowledge.

In her view, what we are really facing is the solutions looking for a problem, the solutions advocated by environmentalists and new business interests involving the decarbonisation of energy supply, taxing carbon and trading it – all major new industries involving much innovation, promising energy security for some (and loss of income to others) as well as much income to middlemen - the financiers, regulators and researchers needed to get global emission counting, trading and taxing off the ground.

She believes that globalisation is the popular philosophy. Global trade, a global fight against terrorism, so why not global action against a threat to the very survival of the human race? This project is particularly attractive to former imperial countries like the UK and, of course, the EU with its efforts at integration. It is for this reason that she believes that the IPCC was primarily interested in mitigation – the global solution. Recently there has been much pressure for adaptation studies and help, from poor countries.

She implies that the IPCC, especially in its more accessible documents produced for policy-makers bends to political influence. As Senator James Inhofe has stated, the IPCC's *Summary for Policymakers* is a political document. It is based on the *Assessment Report*, but is an agreed text, agreed by all the nations involved, and by their ambassadors, not their scientists.

The IPCC accepts, adopts and approves documents, and this difference is important. Approved documents are the most political, and also the most certain about the causation of global warming by fossil fuel emissions. A close study of climate science reveals that there are several other hypotheses trying to explain observed recent (very recent) warming. The so-called consensus theory based on fossil fuel emissions is the only one which also promises solutions. It cannot explain past warm periods, and there have been many, even during recorded history, but before fossil fuels were even used.

Dr Boehmer-Christiansen points out that if global emission control is the solution, implementation puts lucrative tasks in the hands of educated Western elites and commercial institutions. In addition it offers lucrative opportunities to researchers and scientists, again mainly in the West. It puts the West and the developed nations firmly in control of the future of the world. Western companies can provide the expertise and technical solutions to the global community.

Climate science therefore has become politicized. Research institutions through their research agenda and the need to have them funded by governments, have become actors deeply involved in politics. In striving to be policy relevant, these institutions tend to appeal to populist ideologies such as environmental alarmism. Science is generated in order to support policy not the other way around.

The IPCC itself promotes the view that western technologies and markets will save humanity and the implication is that we can control the climate. Environmental alarmism reflects the Zeitgeist of post-industrial society. It is a belief based on selected knowledge and values; values and beliefs that many other human societies do not share (though we are trying hard to persuade them through our foreign policy.) It is a belief that owes at least part of its popularity to its use as justification for both more research and political intervention. Fossil fuel burning is presented as the original sin. We have to seek forgiveness through higher taxes. In fact, emissions continue to rise and targets are expected to be missed. Because all this is an emergency the solutions are put in the hands of bureaucrats - unelected and unaccountable to voters.

In Dr Boehmer-Christiansen's view the reduction in CO_2 emissions was clearly politically attractive to the Thatcher government which wanted to break the power of the miners. At the time the United Kingdom had its own North Sea gas and this was used to fuel power stations which were cleaner than the coal stations they replaced.

For the moment there are few economic incentives for consumers to reduce energy demand, although the costs and availability of energy are likely to be problematic in the foreseeable future. Regulating emissions in the name of saving the planet also lets government take back control over the privatized energy industry.

Is global warming being used as a threat to guide consumers towards energy security? Taxing them may be even more attractive and most likely is the new EU policy. Income tax is always unpopular, but if governments impose taxes on fossil fuels they can raise money and claim they need to do it for the best of green intentions. The higher costs of carbon-containing fuels will benefit their competitors: nuclear power, wind turbines and even biofuels.

Meanwhile, counter arguments are suppressed and scare stories keep people focused and paying more taxes. Climate models cannot be validated. Vested interests in alternative energy are pushing for global warming controls.

Sonja Boehmer-Christiansen paints a picture of global warming as an unbridled opportunity for manipulating public opinion. An opportunity for governments to tax, bureaucrats to control and corporations to profit; all on the back of science that we do not yet fully understand.

REASONED DEBATE

Four sceptics: two views.

Two totally reject the idea of a climate crisis; two question our approach to the solution. Some environmentalists may claim it's too late for talking. They say it's essential to act without delay. I say we must recognise the limits of our knowledge, do what we can but review our options as our knowledge expands.

So far I have tried to take a balanced view and show that:

◊ We have a problem, though no-one is exactly sure of what has caused it. There are different theories and it may take another decade or more to really understand climate and why it changes over time. Extraterrestrial forces may be involved, especially changes in the behaviour of the sun.

◊ We have solutions, but no-one can be sure that they will work or how much they will cost or whether we can afford them, given so many other urgent problems clamouring for investments.

◊ There is no easy answer, and no magic bullet.

That there is a problem and that it is serious, is denied outright by fewer and fewer people. It is, nevertheless, worrying that those who are most concerned and most vociferous seem to believe that all we have to do is to cut carbon emissions and the problem will be over. They ignore the effects of cutting carbon on economic activity – the effects on trade and jobs, and in some places the effects on subsistence and very survival. They ignore the fact that no-one can be sure how things will develop, that climate science is far from comprehensive and certain, even though massive international research is devoted to broadening our knowledge.

The most worrying aspect is that many activists are so confident of the rightness of their beliefs that they want governments to coerce and compel us to do as they say and to tailor our lifestyles as they see fit. And what is equally worrying is that any attempt to question these actions on scientific or libertarian grounds is met with hostility and disbelief – a form of the Bush dictum: if you're not with us, you must be part of the enemy.

This is no argument for doing nothing, but any action we take is like steering a super-tanker. Great effort is needed and it takes a very long time to have any effect. If we have misjudged the winds and the tides our ship will not go where we want it to go. If we have used all our efforts to push it in one direction, which turns out to be the wrong direction, we will end up wrecked on the rocks.

What is the right direction? In spite of all the headlines and hyperbole it is not 'saving the planet'. The planet has been around for millions of years and will survive for millions more. Our objective is not saving the planet; it's saving the human race.

Action now **may** slow down global warming and reduce the effect of climate change for our children and grandchildren. Action now **can** protect vulnerable nations against the present effects of climate change and the effects which are likely to develop and continue for the next century or more. Unfortunately, action now will not prevent sea levels from continuing to rise for another 1,000 years or so, and as sea levels rise some parts of the world will simply disappear. As we have seen, famine and forest fires will destroy lives and livelihoods. Global action is needed to plan, prepare and protect.

Is this enough? Enough for the planet, perhaps, but maybe not enough for the human race. Modern civilisation depends on more energy than has ever been consumed in the whole of human history. Our fossil fuels emit carbon dioxide, the common greenhouse gas, so there must always be a battle between those who want to cut CO_2 and those who want to burn those fuels for the power, energy and comfort they bring.

What if fossil fuels ran out? If they suddenly stopped then CO_2 would stop increasing, but global warming is a fact and would take decades or centuries to reverse. And what would happen to us without fossil fuels? How would we light and heat our homes, get to work, power our PCs, pump fresh water or control disease? Still, fuel is not likely to run out, is it?

That's what the next chapter is about.

Energy for Life!

WILL CLIMATE CHANGE YOUR LIFE?

Supply meets demand?

Some writers say we're addicted to energy; we're junkies – we can't do without it.

I don't find that image very helpful. It paints a picture of someone out of control, desperate for a short term high whatever the cost. If we accuse everyone of behaving like this when we switch on a light, drive a car or take a flight we're unlikely to win their hearts and minds. They won't even listen.

The truth is that we have seen immense technological change from the beginning of the Industrial Revolution through the start of the Information Age to where we are today in the early 21st century.

We have developed mobile telephones, PCs, the Internet, games consoles, digital television, tremendous improvements in medical diagnosis and treatment, advanced avionics, satellite navigation and online banking. All these things, and many others, are an essential part of most of our lives. All of these things depend on the supply of cheap energy which has been developed in parallel with our increasing needs and demands. We use it because we want to: we use it because we can. We use energy to run systems and machines that are more efficient than those that went before. Technology uses energy to create new services, new products, new opportunities and new jobs. Technology uses energy to take over the dull, dirty and dangerous tasks.

If there's a problem with the supply of energy it's not helpful to start by blaming the consumer.

WHERE ARE WE NOW?

The chart below shows that the main sources of energy in the UK are gas, oil and coal. Electricity is not shown because it is not a source of energy but a means of delivering energy. It has to be generated from fossil fuels, biomass or renewable sources like winds, waves and sunshine. Fossil fuels account for just over 90% of the total energy used.

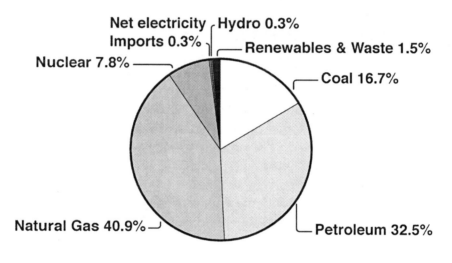

Net electricity Imports 0.3%
Hydro 0.3%
Renewables & Waste 1.5%
Nuclear 7.8%
Coal 16.7%
Natural Gas 40.9%
Petroleum 32.5%

Primary Energy UK 2004
Source: DTI

Nearly three quarters of the oil is used for transport[18] – road vehicles, trains, boats and aircraft. The rest of the energy is generally delivered to the point of use; electricity by the national grid, gas via the pipeline network and heating oil and LPG by road or rail tanker

The two key issues with energy are supply and security.

Supply: is there enough energy to meet our needs?

Security: can we get whatever we want whenever we want it?

••••••••••••••••••••••••

[18]*Department for Transport statistics for 2004;* ***http://www.dft.gov.uk/***

Things have changed in recent years as we shall see.

FOSSIL FUEL FUTURES

Is there an energy supply problem?

Oil

There are still considerable oil reserves in the North Sea, but we have become, once again, net importers of oil. The North Sea cannot meet our demands as North Sea Oil is running out.

The worldwide demand for oil has grown steadily since 1950 and the trend is expected to continue. This means not only that every year we are using up another chunk of the world's declining resources; but every year that chunk is a bigger part of a smaller reserve. As with gas, what the UK cannot source from the North Sea it has to buy on the global market, at market prices. It's not just North Sea oil that's running out, it's happening worldwide.

Running out does not mean that we will wake up tomorrow, next week or in two or five years time and find that the oil wells have run dry. The moment of truth will be when everyone accepts we have reached Peak Oil. Peak Oil is the point at which we have used up half the world's oil and from then on we are truly running out because we can see global reserves coming to an end.

Peak Oil is when the world meets its maximum rate of oil production. Peak Oil means we can no longer satisfy a growing demand. There will be shortages. This rate may be sustained for a few years but after Peak Oil the decline will begin: there will be increasing shortages. No one knows whether the decline will be gentle or catastrophic. We will have used half the planet's oil reserves in just over 100 years. The rate at which we use oil today is so much greater that the second half will be used up much more quickly.

When will Peak Oil occur? Members of the Association for the Study of Peak Oil and Gas (ASPO) base their projections on their personal experiences as oil executives and on the work of geologist M King

Hubbert. A number of oil-producing states, including the UK, have already passed their own oil production peaks.

As long ago as 1956 Hubbert predicted that the maximum rate of oil production in the United States would be reached in 1971. This was strongly disputed by the oil industry, but in the event the peak actually occurred in 1970 and US oil production has declined every year since then. The director of the US Geological Survey, who had led the opposition to Hubbert, was forced to resign.

Using Hubbert's methodology, ASPO believes that Peak Oil for the whole world will occur by 2010, although this may not be recognised immediately by the oil companies, governments or the markets which set the price of oil. By contrast, the general opinion of the oil industry is that Peak Oil is at least ten years away and that total oil reserves will last for at least another 50 to 100 years.

The three factors which determine the timing of Peak Oil are reserves, getting the oil to market and demand.

Reserves

How many billions of barrels of oil do we have remaining on the planet? Well, it depends. The most widely cited source of information is BP's statistical review. This is only a guide, however. It gathers data from sources all over the world; from governments and oil companies (including countries where the government is the oil company.)

Could there be some political bias on some of the figures? The disclaimer on BP's report includes the warning:

The reserves figures shown do not necessarily meet the definitions, guidelines and practices used for determining proved reserves at the company level, for instance those published by the US Securities and Exchange Commission or recommended for the purposes of UK GAAP, nor do they necessarily represent BP's view of proved reserves by country

On the back cover BP excludes itself from commenting further on the figures. This means that the figures are accurate to the best of BP's knowledge. No organisation can examine and verify every item in global reserves. We can only accept the figures provided, and BP's Review is internationally accepted as the best information available.

Let's look at the history. The 1970s oil shock occurred because the OPEC countries, which control the substantial majority of the world's oil production, got together and restricted oil output in order to push the price up. This was almost too successful, so they backed off a bit when they realized what this was doing to Western economies. They did not want to kill the goose that laid the golden egg by destroying their markets in the West.

To maintain the price level the OPEC members agreed that they would restrict output and each country would be allowed to pump oil in proportion to its total reserves. Saudi Arabia, with the largest reserves, pumped the most and the smaller countries pumped correspondingly less. After a while, some of these smaller countries felt they could do with more income. If they just pumped more oil and broke the cartel then they risked a free-for-all where everybody would pump more oil, drive down the price and they would all be worse off. Instead of this, they decided to re-appraise their oil reserves. If they had in fact miscounted and understated their reserves, their production rate was too low. If they found greater reserves then they could pump the same proportion of these greater reserves – more oil and more income. So, during the eighties, the countries re-appraised their reserves.

In 1984 Kuwait found it had overlooked billions of barrels and by the end of the decade six countries, all members of OPEC, had sharply increased their reserves.

Oil reserves can only be increased by new discoveries or enhanced recovery from existing oil fields.

There have been no significant discoveries of oil since the North Sea in the sixties and seventies. The reassessment of Middle East oil reserves, therefore, depended on enhanced recovery.

The Middle Eastern oil wells are some of the oldest in the world and drilling and extraction techniques have advanced immeasurably since they were first opened up. To start with, the oil just gushed out of the ground. Then it had to be pumped. Now engineers can do all sorts of things – inject water or gas into the well to drive the oil out; drill in all directions, including horizontally, to locate pockets of oil; even light fires underground!

Middle East proven oil reserves increased by 53% in the decade from 1985 to 1995, while global reserves increased by 33%. For the 10 years from 1995 to 2005 global reserves rose by 16%, but by 2004/2005 the year-on-year rise was only 0.5%, showing that the rate of increase had almost stopped, at about 1,200 billion barrels. Nevertheless, total reserves were greater than they had been the year before, even allowing for the year's production of 29.6 billion barrels. (2.5%) At the current rate of production, total reserves will last for 40 years. Demand, however, is growing and if it is allowed to continue to grow the reserves will run out much sooner.

Until now, for every barrel that has been taken out of the ground, more than one new barrel of reserves has been found. This does not mean new discoveries– there have been few significant finds for some years. It means largely that existing reserves have been re-appraised, but even so, we are now very close to the point where production exceeds new reserves. That is Peak Oil; the point at which oil will truly be running out. That will not be the end of oil, but it will definitely be the end of cheap oil.

Getting the oil to market

The second factor which will determine when Peak Oil comes is the capacity of the distribution system – the wellheads, the pipelines, the tankers, the refineries. Since more and more oil has to come from very remote areas, pipelines are getting longer and longer. They often run through earthquake zones and politically unstable regions. They are an ideal target for terrorists: attacks on Iraq's pipelines and oil infrastructure mean that the country's oil exports are still significantly less than they were before the invasion.

BP heads a consortium building a 1768km pipeline from Azerbaijan through Georgia to Ceyhan in Turkey. Construction is costing some $3,000,000,000. Imagine the difficulty and the cost of the security and maintenance of an installation like that! It takes $600,000,000 worth of oil just to fill it up. Such a pipeline can only be justified if the size of the field is big enough. Small pockets of oil in remote places will never be worth recovering even if the price doubles or triples. Some reserves are therefore unusable 'stranded assets'.

Demand

The third part of the equation is the demand for oil. The faster demand grows; the sooner Peak Oil will be reached. Demand is growing steadily in the USA and Europe. Demand is growing very rapidly in India and in China, soon to become the world's second largest economy.

What happens when we reach Peak Oil, the point where we still have oil, but there is no longer enough to go round? Markets react to a shortage by raising prices.

The oil price went up in the seventies when OPEC reduced supply. The price went up at the time of the Yom Kippur war, at the time of the first Gulf War and the invasion of Kuwait by Iraq. The price went up in autumn of 2005 when Hurricane Katrina destroyed refinery capacity and disrupted production in the southern states of the USA. Prices have continued to climb in the face of difficulties in Iraq and problems with Iran.

The markets have raised prices to protect themselves against the risk of short term shortages to a level just short of $80.00 a barrel, compared to the long-term average of just $21.00 a barrel. When the perceived threats recede, the price falls back.

After Peak Oil there will never again be enough oil to go round. So far markets have refused to accept that we have a problem that will have consequences on any timescale that is relevant to them. When the markets eventually realize that we have a permanent problem, a problem which will prevent the constant increase in economic growth, we will see a whole new situation. This is where prediction gets difficult. This is a situation we have never seen before.

In the past, short term oil price increases have brought recession to Western economies. There are those who predict that Peak Oil will bring a stock market crash as bad as 1929, followed by depression worse than the 1930s. This is where most people switch off, because the prospect is just too extreme.

If we believed the hype about everything from the millennium bug to bird flu we would all stay in bed. If the solution to our problems means not driving cars, not heating our houses, not taking cheap flights to the sun, most of us are going to do all those things while we can, and not worry about it until something stops us. Of course by then it may be too late, but on the other hand, like the best scare stories, it may never happen. As a recent letter to a newspaper said, "I'm not going to stop flying to Australia until everybody else stops driving cars and they all start walking to work."

But maybe all this is overplayed. If one source of oil runs out maybe there are others.

Unconventional oil

Whatever we do, there is no doubt that oil is going to be more expensive. There are very few places where oil simply gushes out of the ground any more. It has to be pumped out or gas or water has to be pumped into the oil well to force it out. Increasingly, oil is being produced from more and more inhospitable places; either remote, unstable regions or offshore, in increasingly distant and dangerous waters. All these factors increase the cost of production and increase the costs that consumers must pay.

As these prices rise it makes it economic to look at unconventional sources of oil. These are not cheap options but they are alternatives when the extraction of conventional oil gets more difficult and expensive.

Tar sands

Tar sands, or oil sands as they are known in Canada, are found all over the world, but mainly in Alberta, Canada and along the north bank of the Orinoco River in Venezuela. The Canadian government estimates that the potential reserves are second only to the reserves in Saudi Arabia, in other words, they are the second largest in the whole world.[19] The reserves in Venezuela are estimated at about 1.2 trillion barrels, which is about the same size as Canada's reserves and more than all the oil that has ever been used. With reserves like this, how can we possibly have a supply problem for years and years to come? The answer is in the costs and the difficulties of production.

Tar sands are mineral deposits which contain about 10-12% bitumen, a thick tarry substance. The remainder is 80-85% mineral matter – including sand and clays – and 4-6% water.

Bitumen, like crude oil, has to be refined to yield the petroleum products which are required by transport or industry, but it needs some additional upgrading before it can be refined. It is also far too thick and sticky to be transported by pipeline unless it can be diluted with gas or light oil. This is a particular problem in Canada, where the winter temperatures can go down to -40°, making bitumen an immobile solid.

Light oil or gas must be delivered to the tar pits so that the bitumen can be diluted and pumped out, because the alternative, building a refinery next to the tar pit, is just too expensive. Transporting the bitumen is a cost, but the major cost in exploiting tar sands is in extracting the bitumen from the sand, clay and water which make up the rest of the deposits.

Some tar sands are found near the surface of the earth and extraction is by opencast mining. Such mines are enormous operations. For every ton of bitumen about ten tons of tar sands need to be dug up. To squeeze out the oil needs tonnes of hot water. After the sand is excavated, the material is crushed and then treated with 40°C

....

[19] **http://www.energy.gov.ab.ca** *The Alberta Government Energy website*

water and caustic soda to turn it into slurry. It is then pumped to an extraction plant where even hotter water is added. The whole mixture is agitated and the bitumen skimmed from the top. Every cubic metre (1,000 litres) of oil produced requires between 2000 litres and 4500 litres of water. A million barrels a day of oil production translates into roughly 2 to 4.5 million barrels of water a day.

Major tar sand reserves lie deep underground. It is not possible to dig out the enormous quantities of tar sand needed for processing, so techniques for 'in situ' production are used – both involving steam.

With **Cyclic Steam Stimulation** steam is injected into a well for several days at a time to heat the sand and cause the tar to flow. It can then be pumped to the surface. When it stops, more steam is injected.

Steam Assisted Gravity Drainage uses constant steam injection to the top of the well and pumping from the bottom of the well. The idea is that the steam will raise the temperature and the condensate will wash the tar to the bottom of the well, so it can be pumped to the surface.

The investment in setting up oil extraction from tar sands is substantial. It has been estimated that it costs $15-$20 per barrel just to get the oil out of the ground, compared with a cost nearer $2 per barrel for conventional oil. That is one limit to the output from the tar pits.

Another limitation is the cost of energy to raise the steam to release the oil, whether in extraction plants or deep underground. Natural gas used to be a cheap by-product of oil production, but now it has a value in its own right. It has been suggested that nuclear power stations could be built and the waste hot water used for the extraction process. Quite apart from the difficulty of building a nuclear power station in such a remote area, nothing could be done quickly enough to affect production levels for the next 10 to 15 years.

Water is another constraint. The main Canadian tar pits are in the valley of the Athabasca River, but there are already concerns that the vast quantities of water used in extraction are affecting other users, notably farmers. The plan is to treble output by 2015: this implies a need for three times as much water.

Such expansion will also need investment in pipelines to bring the oil to market as well as major investment in extraction plants and infrastructure. A major constraint is believed to be a shortage of qualified engineers ready and able to go to work in this remote area. If they can be found, the money they will demand will go towards pushing up the cost of this very expensive oil.

And then there are environmental concerns. There may not be enough water, but every litre of water that is used must be cleaned or recycled.

Can we be sure that steam injected into tar pits will not condense and contaminate the water table? Something must be done with the residues from the extraction plants at the open-cast sites.

Concerned citizens are vociferous about the danger to Canada's forests and have recently issued a report entitled "Death by a Thousand Cuts: The Impacts of In Situ Oil Sands Development on Alberta's Boreal Forest"[20] They complain that with no public discussion, and without a plan in place to protect the forest, the Alberta Government has already leased out over 35,000 km² of forest for deep oil sands development - an area nearly twice the size of Wales.

Trebling Canada's output of heavy oil by 2015 means producing 3 million barrels per day instead of the 1.1mbd produced in 2005 limited pipeline capacity.[21] Oil production in Venezuela for 2005 was estimated to be 2.25mbd but there are doubts whether sufficient investment is available in that country even to maintain that level of output. Putting these figures in context; total global production was 81mbd in 2005, of which 25mbd came from the Middle East.

· · · · · · · · · · · · · · · · · · · ·

[20]Pembina Institute and the Canadian Parks and Wilderness Society
[21]National Energy Board of Canada

The US Energy Information Administration projects demand of 98mbd in 2015, so even though Canada and Venezuela possess these colossal reserves, their contribution to global consumption will still only be between 3% and 5%.

When we reach Peak Oil, tar sands are clearly not the answer.

Shale

While the United States has the largest share of the world's oil shale resources, the only producers of oil from shale are Brazil, Australia and Estonia. Reserves are potentially enormous, as big as tar sands, as big as all the oil we have ever used, but production difficulties are enormous as well.

Strictly, oil shale does not contain oil, but it is a type of rock from which a petroleum substitute can be distilled. Oil shale can also be burnt like coal, though it leaves substantial quantities of ash. Distilling petroleum from shale is similar to the extraction process for tar sands: it needs heat and water.

Shale can be treated in a retort or in situ by pumping water and heat underground. The problem with using a retort is that shale leaves a residue which is 20% greater in volume than the shale that was mined in the first place, producing a disposal problem as well as the problems of energy inputs, water supply and recycling.

Several organisations in the US have attempted to produce oil from shale commercially, particularly during the oil shock of the 1970s, but none has been successful. Shell, the oil company, is researching methods of in-situ extraction, and 'expects to reach a commercial decision near the end of this decade.'

Oil from shale is no short-term solution to Peak Oil.

Unconventional oil – tar sands or shale – cannot make up for the decline in conventional oil. Even if problems of energy inputs, pollution and water demand could be solved, oil output would supply only a minor fraction of current usage. Unconventional oil is still a fossil fuel and using it still releases CO_2.

Is gas the answer to the shortage?

Gas

In 1970 natural gas accounted for less than 6% of the energy used in the UK. By 2004 that proportion had increased to 40% and gas had become the UK's largest source of energy – more important even than oil as a fuel. The security of our gas supply is, therefore, very important to our businesses, industries and everyone's lifestyle. The UK is in fact the largest market for gas in Europe.

Gas is probably the most versatile fuel we have. We use it to generate 34% of our electricity, it's the most popular fuel for central heating, it's widely used for cooking and we can use it to drive vehicles.

Gas is cleaner than coal or oil. It emits less CO_2, particulates and soot. Best of all, vast gas fields were found along with the oil in the North Sea. In the 1970's Britain gave up town gas made from coal and converted to North Sea gas. In the 80's and 90's Britain closed coal-fired power stations and built cheaper and cleaner gas-fired stations. The change to gas has been very significant in Britain's progress towards the Kyoto targets for the reduction of CO_2 emissions.

As long as gas flowed from the North Sea the UK was in a strong position. However, while total global reserves of gas are estimated at 179.83 trillion cubic metres, the UK has only 0.3% of these, and they will last only six years at the current rate of production.[22] UK gas production for 2005 was already down by 8.1% from the previous year. Consumption fell by 2.2% but still exceeded production: Britain is now a net gas importer. There will be a steep rise in imports as North Sea production declines and the government expects imports to cover 80% - 90% of the requirements by 2020.[23]

Globally, ASPO expects Peak Gas to be reached around 2015. Global reserves amount to enough gas to last 65 years at current rates of consumption – less, of course, if demand doubles by 2020 as

....

[22]Figures in this section from BP's Statistical Review of World Energy 2006 unless otherwise stated.
[23]The Energy Challenge – DTI, 2006

predicted. 40% of these reserves lie in the Middle East and 32% in the former Soviet Union.

At present the UK sources gas by pipeline mainly from Norway and the Netherlands. The Langeled pipeline from Norway to Easington, Yorkshire, was opened in October 2006 with sufficient capacity to supply 20% of the UK's gas needs. We also import LPG (liquid petroleum gas) by refrigerated tanker from Algeria and other Middle Eastern countries.

Russia is likely to be the major player as far as our gas supplies are concerned in the future, but Russia has already shown how it can use its power to control the market. In January 2006 it demanded a price increase from Ukraine and when Ukraine refused, Russia cut off supplies. In fact, since the pipelines that supply Ukraine cross the country to supply other countries, Ukraine simply cut off supplies to Slovakia and Hungary.

The dispute was eventually settled. Ukraine had been paying extremely low prices for its gas, to the extent that consumers would open windows in mid-winter when rooms got too hot rather than turning down the heat, and the new agreement set prices at a more realistic level. However, the episode shows that countries can suffer supply disruptions even if they have no dispute with their suppliers. It is not known whether Slovakia and Hungary were ever compensated for the shortages they suffered during the Russia/Ukraine dispute.

The UK is at the far end of the European pipeline network, and so is at greatest risk from problems elsewhere in Europe – disputes, breakdowns or terrorism. The government recognises this and is encouraging the private sector to make investment in storage facilities, pipelines and additional LPG terminals. If there a significant delay to the construction of this infrastructure the lack of any buffer between demand and supply could lead to relatively high and volatile prices, which could have a considerable impact on the economy.

The DTI's 2006 Energy Review states that if new infrastructure is not forthcoming or is delayed, there is a risk of price rises costing consumers hundreds of millions of pounds. If demand rises and supply is restricted then prices will rise. For example, a 1p/therm increase in price on a winter day adds approximately £1 million to the wholesale cost of gas; over a winter this might equate to some £200 million, and all this is passed on to consumers. As a fuel, gas is more complex and more costly to store than coal or oil and there are currently no international arrangements to manage disruption to supplies, unlike in oil.[24]

Although the government considers there is a minimal probability of unplanned gas cuts before 2014, provided the infrastructure is completed in time, it recognises that the consequences of the loss of the gas supply, even for a short time, would be very serious.

If gas is cut off, safety regulations demand that every appliance must be checked before the supply is restored. This is because if air gets into the system it can cause catastrophic explosions. For this reason, if cuts are necessary they will involve industrial users including electricity generators long before domestic consumers are affected. The irony is that if electricity generators are affected there will be domestic power cuts, and since most gas heating systems have electric control systems they won't work either.

Apart from short-term disruptions to supplies, there will be growing long term pressures. Gas production is declining in North America. China, rapidly industrialising and with 20% of the world's population, has only 1.3% of global gas reserves – less than Australia. In fact China has a major contract with Australia for the supply of gas. All countries must look to world markets for their requirements and the markets will allocate gas supplies. Only those prepared to pay the asking price will be successful.

In recent months we have seen gas prices to domestic consumers in the UK rise by 20% and more. The rises are set to continue. If they continue to rise by 20% per annum they will double every three to four years.

. . . .

[24]*The Energy Challenge, Energy Review Report 2006, Department of Trade and Industry, Crown copyright*

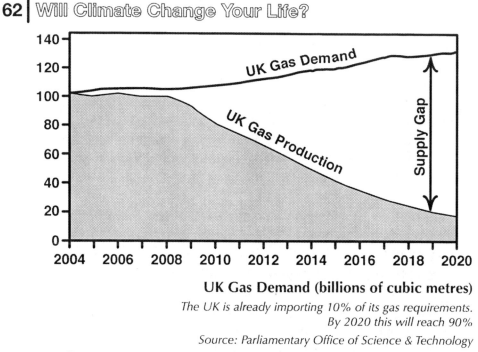

UK Gas Demand (billions of cubic metres)

The UK is already importing 10% of its gas requirements.
By 2020 this will reach 90%

Source: Parliamentary Office of Science & Technology

Coal

Remember coal? Maybe not – you don't see it so much these days, but it is actually still a major source of energy.

Before gas central heating became popular in the 1970's every house had a hearth, or several. Every house had its coal-hole, coal-shed or cellar, but by 2005 coal used directly accounted for only 1% of UK energy. Indirectly though, we used 39.1 million tonnes oil equivalent (mtoe), mainly for electricity generation.

Coal accounted for 38% of electricity generation in 2004 compared with 34% from gas. This is a dramatic turnaround from 1980, when coal accounted for 73% and gas for just 1%. This change emphasises the change in Britain's energy security which has occurred in those 24 years.

In 1980 we imported 5% of our coal. In 2004 we imported 59%,[25] mostly from South Africa, Russia and Australia, but some from as far away as China and Vietnam.[26] Electricity prices are now affected by

....

[25]*Department of Trade and Industry*
[26]*Digest of UK Energy Statistics 2005*

the strength of the £ sterling against the rouble, the Australian dollar and the South African rand!

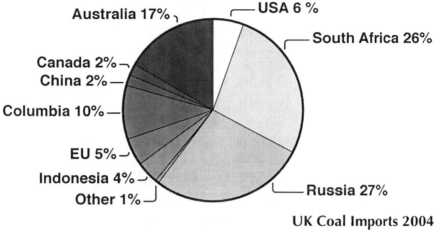

UK Coal Imports 2004

While coal still comes from Yorkshire, we now import it from all over the world.
Source: UK Digest of Energy Statistics

According to BP's *Statistical Review of World Energy 2006* the UK has reserves of 220 million tonnes of coal, sufficient for a further 11 years of production. BP makes it clear that the reserves shown in its report relate to coal which can be economically recovered using current production methods. Other estimates, which include significant deposits beneath the North Sea and other reserves that cannot be mined conventionally, put the UK's total reserves at anything between 1000m and 2500m tonnes.

Clean Coal

Coal has a bad name because it is not a clean fuel and mining is dirty and dangerous. The main concern is that coal produces more CO_2 than gas or oil, as well as many other pollutants, depending on the type of coal used. Much can be done to overcome this, from the design of the combustion process to 'scrubbing' the flue gases.

New power stations which have been built with these designs are much cleaner than older ones. Old stations can be modified, but the greater cost of retro-fitting is often not economic. Anything that cleans up emissions needs energy and, therefore, reduces the efficiency of the station to some extent.

New designs and modifications have limited effect on the carbon dioxide created by power stations – most of it goes up the chimney. Currently this amounts to some 50 million tonnes of carbon (mtC) per year, or 30% of all CO_2 produced in the UK from all sources. With the government committed to reducing CO_2 emissions by 60% by 2050, power station emissions will have to be controlled. The favoured solution is carbon capture.

Theoretically, carbon capture and storage will reduce the CO_2 released by a fossil-fuel power station to the atmosphere by 80%-90%. The CO_2 is collected and piped to a disposal point. In Canada CO_2 is injected into an oil field to improve oil recovery. In Norway CO_2 is injected into a deep saline aquifer. It has also been suggested that CO_2 could be stored on the deep seabed. The tremendous pressures would transform it to a solid and imprison it there indefinitely. However, no commercial scheme involving electricity generation and carbon capture has yet been constructed anywhere in the world. Any scheme is going to involve extensive infrastructure – gas separators, compressors, pipelines, selection and development of suitable storage sites.

The next fuel source to explore is Underground Coal Gasification (UCG). This is a process which releases CO_2, but at the same time it creates underground cavities for storage of CO_2, both from the UCG process itself and possibly from coal-fired power stations as well.

All this spells cost – costs which we all must pay, either as consumers or as taxpayers. It is another solution which will not happen in the short term. Further research is needed, probably a pilot plant and then construction of a full-scale unit. But no-one will even consider it unless the financial arguments make sense.

Unconventional Coal[27]

If we can exploit 2,500 million tonnes of reserves instead of the 220 million tonnes available to conventional techniques, we will have enough for more than a century. There are several possibilities, though much more research and development is needed.

....

[27] *The Coal Authority* **http://www.coal.gov.uk**

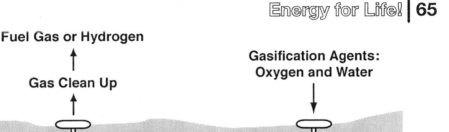

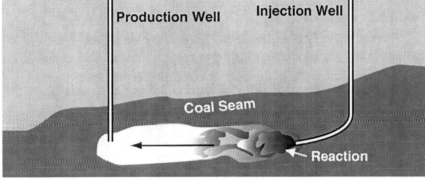

Underground Coal Gasification

Underground coal gasification (UCG)

Instead of mining the coal it can be turned into gas underground. This is done by drilling boreholes into the seam. Oxygen and steam are injected to start partial combustion of the coal to drive off the gas. This is extracted from another borehole where the CO_2 can be stripped out leaving a clean fuel to be piped to consumers or used to generate electricity from gas turbines.

The CO_2 has to be captured, but long pipelines are not necessary as it can be re-injected into spent parts of the mine. In addition, UCG leaves none of the spoil heaps associated with mining and there is only limited infrastructure on the surface.

UCG has been used in the Soviet Union and in Uzbekistan and there is a trial operation in Australia. There were small-scale trials in the UK in the 1950's, but beyond that there are no commercial operations and there have always been problems with drilling successful boreholes and producing gas of consistent quality.

A £12m research programme was launched in the UK with the objective of commissioning a pilot plant in 2005. The initial planned trial did not take place, but other sites were considered, and one area in particular on the Firth of Forth was the subject of a feasibility study (2004-2005). The next phase of the study, which could involve exploration and drilling, was still under consideration at the end of 2006.

Meanwhile, other countries are pressing ahead with commercial UCG, with trials in China, South Africa and Australia and a variety of feasibility studies elsewhere including the USA. UCG in deep coal seams is particularly attractive as a low or zero emission source of energy because CO_2 capture is less costly than alternative (because of the high partial pressure) and the void has the potential to act as a permanent store of CO_2.

Coal mine methane (CMM)

Since methane comes from decomposing organic matter and since coal was originally pre-historic plants, methane – the dreaded 'fire-damp' - is commonly found in coal mines. It still leaks from abandoned mines, and it is a problem in working mines so ventilation is very carefully managed to extract the methane safely.

A number of installations have been put in place to collect and utilise the gas, either for power generation or for delivery to the gas grid. Production comes from both working and abandoned mines, and some plants generate up to 10 MW of electricity.

Coal bed methane (CBM)

Methane trapped within coal seams can be released by pumping out the water to reduce the pressure and then collecting the gas. The process can prevent methane from simply escaping to the atmosphere, but the water extracted is frequently strongly saline. There are no commercial CBM projects in the UK, but there is a major extraction facility in the Powder River Basin in the US. There is strong controversy about the disposal of the extracted water because of its salinity. Salt water is no good for drinking, kills plants and animals and poisons agricultural land. Unless it can be safely disposed of – and we're talking of millions of gallons – the project may be halted.

THE NUCLEAR OPTION

The first commercial nuclear power station in the world was opened at Calder Hall in the North of England in 1956. With no computers and limited electronics it is remarkable that the station worked successfully and safely delivered up to 196 megawatts of power for more than 40 years. After all, in nineteen-fifties Britain most trains were drawn by steam engines, many farmers still used horses and the normal way to cross the Atlantic was by ship. You needed little more than a screwdriver and a spanner, and sometimes a hammer, to maintain your car, but here was a process that required the minute manipulation of atoms themselves. Nevertheless, the station remained operational until it was taken out of service in 2003.

Whereas a coal-fired power station of the same output as Calder Hall needed 400,000 tons of coal per year, a nuclear power station used just over 30 tonnes of uranium for the same annual output.[28]

'Electricity too cheap to meter!' screamed the popular press, hoping for an era of unlimited free power. Certainly the fuel cost was low, but power stations, whether coal or nuclear, cost millions to build and the national grid and local supply networks were still needed to get the power to the consumer. And nuclear power stations leave radioactive waste. Not just spent fuel, but parts and components which have been exposed to radiation and will remain radioactive for thousands of years. All of these need to be accounted for and safely stored indefinitely.

• • • • • • • • • • • • • • • • • • • •

[28]*http://www.opendemocracy.net/conflict-climate_change_debate/2587.jsp#four*
A Nuclear Power Primer by Jan Willem Storm van Leeuwen He goes on to point out that mining granite yields about 0.0004% uranium, and a further 50% is lost in processing before it can be used in a power station. 500,000 tonnes of rock must therefore be mined for each tonne of uranium fuel. Mining is not the only process. Uranium cannot be practically extracted from seawater, but http://www.uic.com.au/nip04.htm suggests that significant supplies are available from redundant warheads. http://www.world-nuclear.org/info/inf23.htm Principal producers of uranium are Canada and Australia.

When a nuclear plant comes to the end of its useful life, as Calder Hall has done, (also Dungeness A and Sizewell A on 31st December 2006), it needs to be decommissioned and dismantled. It takes a specialist staff, rigidly protected against contamination, many years to take a power station apart, separate everything that remains, reprocess it and prepare for storage. Storage facilities must be secured and guarded for all our lives and for generations to come. All this adds immeasurably to the cost of power once thought 'too cheap to meter'.

All nuclear power stations currently in operation work on the same basic principle of nuclear fission. A controlled chain reaction splits atoms which stimulate other atoms to split, and this releases heat.

In a coal-fired station coal heats water and makes steam to drive turbines. A nuclear reactor heats water and drives turbines in the same way. The difference is that the nuclear station emits no CO_2. It creates no sulphur dioxide, carbon monoxide or anything that causes acid rain. The reaction produces radiation, but all of this is contained within the reactor.

Views on nuclear power are strong and violently contested – whichever view they support. James Lovelock, originator of *the Gaia theory of whole earth science*, believes that nuclear power is the only option for energy as it emits no climate-damaging greenhouse gases. Many people oppose him on this: leading environmental groups will have nothing more to do with him.

Those that oppose nuclear power point first at the enormous whole-life cost. There is no doubt that the true cost of nuclear power must include the costs of decommissioning and storage as well as the cost of construction and operation. Nuclear power is also painted by many as the most dangerous source of electricity. They cite the accident at Three Mile Island, the disaster at Chernobyl and they point out that if terrorists crashed an aircraft onto a nuclear power station the results would be catastrophic.

Let's look at these in turn.

Nuclear accident

Three Mile Island

The accident at the Three Mile Island Unit 2 (TMI-2) nuclear power plant near Middletown, Pennsylvania, on March 28, 1979, was the most serious in U.S. commercial nuclear power plant operating history, even though it led to no deaths or injuries to plant workers or members of the nearby community. Nevertheless it brought about sweeping changes involving emergency response planning, reactor operator training, human factors engineering, radiation protection, and many other areas of nuclear power plant operations. It also caused the U.S. Nuclear Regulatory Commission (NRC) to tighten and heighten its regulatory oversight.

Although the TMI-2 plant suffered a severe core meltdown, the most dangerous kind of nuclear power accident, it did not produce the worst-case consequences that reactor experts had long feared. In a worst-case accident, the melting of nuclear fuel would lead to a breach of the walls of the containment building and release massive quantities of radiation to the environment, but this did not occur as a result of the accident.

Detailed studies of the radiological consequences of the accident were conducted by US government agencies including the NRC and the Environmental Protection Agency, several independent studies were also conducted. It was estimated that the average dose to about 2 million people in the area was only about 1 millirem. To put this into context, exposure from a full set of chest x rays is about 6millirem and, compared to the natural radioactive background dose of about 100-125 millirem per year for the area, the collective dose to the community from the accident was very small.

In the months following the accident thousands of environmental samples of air, water, milk, vegetation, soil, and foodstuffs were collected by various groups monitoring the area. Very low levels of radionuclides could be attributed to releases from the accident. However, comprehensive investigations and assessments by several well-respected organizations have concluded that in spite of serious damage to the reactor, most of the radiation was contained and that the actual release had negligible effects on the physical health of individuals or the environment.

The Chernobyl Disaster[29]

The catastrophic Chernobyl accident in the former Soviet Union, in 1986, was by far the most severe nuclear reactor accident to occur in any country. It is widely believed an accident of that type could not occur in U.S.-designed plants.

Horror at the Chernobyl disaster halted virtually all nuclear power development across the world, although with hindsight it was not nearly as dangerous as first thought

The accident in reactor no. 4 at the Chernobyl nuclear power station on the western border of the Ukraine, which took place on 26th April 1986, was the worst that has ever occurred.

The operating team was undertaking a test and, in order to prevent it being interrupted, they had deliberately switched off the safety systems. When things began to go wrong and a sudden and unexpected power surge developed, there was no emergency shutdown system to contain it. The reactor went out of control and a violent explosion blew the 1000-tonne sealing cap off the reactor building. The fuel rods melted, the graphite in the core ignited and the resulting meltdown released 100 times more radiation into the atmosphere than was produced by the atom bombs dropped over Hiroshima and Nagasaki.

Surrounding areas were contaminated by radioactive products, and the winds carried much of the material across Scandinavia and Western Europe. Indeed, soaring levels of radioactivity made the Scandinavians aware of the disaster some time before the Soviet Union admitted that anything had gone wrong.

Emergency teams moved in to control the situation, and eventually the fire was extinguished and the remains of the reactor were stabilised and encased in concrete. Some 200,000 people were

....
[29]**www.chernobyl.info** The international communications platform on the long term consequences of the Chernobyl disaster

evacuated from the immediate area and an area extending for a radius of 30km around the plant was classified as an exclusion zone. It remains so today, but nevertheless around 5.5 million people, including 1 million children, still live within this the region.[30]

In 1986 it was predicted that thousands of people would die as a result of exposure to radiation. The intensity of radiation declines with the distance from the source, so the emergency teams and those working at the station received radiation doses far more intense than those living further away.

It was also assumed that there was a direct relationship between the level of radiation received and the risk of health problems, principally cancer. In other words, however low the radiation dose they received, individuals would still have an increased risk of cancer to some extent. It now seems that there is a threshold level below which this is not true. Some scientists even believe that exposure to radiation at very low levels can be beneficial.

The health of the population around the reactor, as well as the health of the thousands of emergency workers from Russia, has been carefully studied over the 20 years since the accident. The number of fatalities occurring up to 2005 and directly attributable to the accident is estimated by the Chernobyl Forum[31] to be 50. These are emergency workers who received very high exposure to radiation in the course of the clean-up operations.

A report by the World Health Organisation[32] concludes that by 2004, 28 of the emergency workers involved in the immediate containment operation had died of Acute Radiation Sickness as well as a further 19 who died of other causes. A 4.6% increase in death-rates due to radiation-induced illness was observed among the Russian emergency workers in the 12 years following the accident.

No radiation-induced mortality was detected in the local population, though there have been extensive health problems, the most

• •

[30]**http://www.chernobyl.info/** - see geographical location page on this site.
[31]**http://www.iaea.org** Document DC 06092005
[32]Health Effects of the Chernobyl Accident and Special Health Care Programmes – Report of the UN Chernobyl Forum Expert Group "Health" – Geneva 2006

prominent of which has been a dramatic increase in thyroid cancer among children. This is generally accepted to result from the contamination, but it is important to note that the thyroid cancer has been successfully treated in more than 98% of cases.

Comparative studies with populations in Belarus, Russia and Poland have shown that in the years after the accident there were pressures on health and mortality in all these countries arising out of stress from the break-up of the Soviet Union. There were additional pressures on those evacuated from the Chernobyl region. They were shunned by those who feared they were radioactive, criticised because they were offered better housing in their new locations than the original population had enjoyed. They were rejected by possible marriage-partners for fear of genetic mutation (groundlessly). All this led to stress, alcoholism and other health problems.

Counting the cost

Nobody should minimise the effects of Chernobyl. Even though there were less deaths than might have been predicted; even though the health consequences were not as serious as were suspected, the fact is that the accident caused illness, distress and anxiety to hundreds of thousands people in Ukraine, Russia, Belarus and right across Europe. Every precaution must be taken to ensure that this sort of incident never occurs again.

Does this mean that we should never build another nuclear power station? Leaving aside the issues of cost and whether or not nuclear power is emission free, the first question is whether or not nuclear power is safe.

Without being pedantic, nothing is totally safe. Aircraft crash, motorists have accidents, hospital operations fail. We are aware of all these risks and we do everything we can to minimise them. At a certain level they are acceptable; we recognise that we must accept them unless we are prepared to forgo the benefits of air travel, the convenience of the car or the prospect of relief from pain or disease.

Like all these things, nuclear power is not totally safe, but what are the true risks? How likely is it that a Chernobyl accident could happen again?

The first cause of the Chernobyl accident was human error. Switching off the safety systems was a reckless act. Arguably the operators were not sufficiently trained. For example, they did not know that in certain circumstances lowering the fuel rods to stop the reaction could have the opposite effect with this particular design of reactor. In fact a similar situation had arisen in a reactor in Latvia and had been successfully contained. No-one at Chernobyl knew about this.

Secondly, the design of the reactor and of the containment building was not of a type which would be permitted now. Construction of the Chernobyl plant was started in 1972. Our knowledge of nuclear power, and particularly of computers and control systems, has developed immeasurably since then.

A third factor which affected the level of casualties resulting from the incident was the level of secrecy which existed in the Soviet Union at that time. If the authorities had been prepared to admit earlier that a major incident had occurred evacuation could have started more quickly and the exposure of the population to radiation might have been reduced.

The emergency workers who were first on the scene are the true heroes: they contained the incident but were exposed to massive radiation and many suffered and died from Acute Radiation Sickness within weeks. Better contingency planning may have saved them from this.

In summary, there are many reasons why an incident such as the Chernobyl explosion would not happen again. There are reasons, too, why such an incident occurring now could be better contained and would have lesser consequences. There will always be a risk from nuclear power. In the UK we accept 3,000 deaths each and every year as an acceptable price for road transport. The social cost of nuclear electricity is tiny by comparison.

Terrorist attack

What if terrorists crashed an aircraft onto a nuclear plant? It depends on the type of aircraft, the type of nuclear plant and to some extent on the weather. In the worst case a direct hit could perhaps release radioactive material into the atmosphere and winds in the right direction could blow it across centres of population. Rain could bring this material to earth and contaminate the soil. Even if this contamination was –significant – and the Chernobyl experience seems to indicate that this is unlikely – is this an argument against nuclear power stations or against aircraft?

Think of the damage caused by an aircraft crashing into a central London railway terminus at rush hour, or into half a dozen planes lined up at Heathrow or into a fuel depot like the one at Buncefield.[33] As long as aircraft fly and terrorists exist, all these risks are possible. As long as we want to fly we ensure that these risks are minimized. Everyone thinks that the radioactivity factor makes destroying a nuclear power station so much more dangerous. The facts prove otherwise. The nuclear industry takes stringent safety precautions, just as airlines do.

Nuclear waste

Although the waste from a nuclear reactor remains radioactive for hundreds of years the level of radioactivity declines fairly rapidly and the amount of protection needed becomes less and less. James Lovelock, a distinguished scientists and Fellow of the Royal Society, (so he should understand these things!) has said that he would be happy to store the waste from a nuclear power station in his back garden. Given the proper concrete container to keep it in he would have no qualms about his grandchildren playing on top of it and he would use the residual heat to heat his house.

Many people would be horrified at the thought of a source of radioactivity so close to home, but some radiation is after all natural.

....

[33]*In December 2005 an explosion and fire at the Buncefield oil depot near Hemel Hempstead caused the destruction of a nearby industrial estate. Since this occurred on a Sunday morning there were no fatalities among the 3,000 people who normally worked there. The fire took several days to extinguish and was classed as the largest blaze in Europe since the Second World War.*

In parts of Devon and Cornwall, where there is a high concentration of granite, there are natural background levels of radiation far in excess of EU 'norms' and there always have been. Nobody suggests that these places should be evacuated.[34]

Studies following the Chernobyl explosion suggest that low-level radiation is not as dangerous as once thought: some of the data implies that exposure to low-level radiation is not dangerous at all.

NUCLEAR FUSION – THE PHILOSOPHER'S STONE?

Ancient mystics were constantly seeking for the philosopher's stone – something that would transform lead into gold. Modern physicists are on the track of a method of turning sea-water and rock into limitless electricity.

Existing nuclear power stations use nuclear fission. Many people hope that nuclear fusion, a process with all the benefits of fission and few of its problems, will soon replace it. Fusion is a process which is relatively clean and radiation free. The radiation that fusion creates lasts only 50 years or so, compared with the thousands of years of radioactivity created by nuclear fission.

Fusion is the process which goes on in the interior of the sun, producing temperatures in excess of 15,000,000°C. Instead of atoms breaking up, the heat is generated by atoms coming together. The process is as clean as nuclear fission with no greenhouse gases or other toxic emissions, and with far less radiation.

The energy gained from a fusion reaction is enormous. To illustrate, 10 grams of Deuterium (which can be extracted from 500 litres of water) and 15g of Tritium (produced from 30g of Lithium) reacting in a fusion power plant would produce enough energy for the lifetime electricity needs of an average person in an industrialised country.[35]

. . . .

[34]*http://www.hse.gov.uk/radiation/ionising/radon.htm* Though it is recognised that concentrations of radioactive radon gas can build up in enclosed spaces, including homes, and are a health hazard. The British government advises preventive measures. Lovelock's bunker would of course be out of doors!

[35]*http://www.jet.efda.org/*

Research into reproducing nuclear fusion on earth has been going on since the 1940s.[36] The United Kingdom's Atomic Energy Authority has been heavily involved in this through its START, JET and MAST projects.

Although fusion on earth is theoretically possible, much lower pressures on earth require much higher temperatures than on the sun. The challenge is to trigger a reaction at 150,000,000°C, and the main part of the challenge is to find a container that will confine the reaction at such temperatures. In fact there is no material on earth that will withstand such heat so the plasma that is used is confined by magnetic fields. In turn, vast amounts of electricity are needed to maintain these fields.

The JET project has generated electricity, but only 70% of the power required to maintain the magnetic fields. The next project (ITER) is planned to generate ten times as much energy as needed for the magnetic field, thus producing a surplus for output to the distribution grid.

ITER – the International Thermonuclear Experimental Reactor – is being built jointly by the European Union, Japan, China, South Korea, Russia and the USA at a cost of some €5,000,000,000. It is under construction at Cadarache in the South of France and will be operational in 2015. If all goes well a demonstration fusion power station will be supplying electricity to the grid from 2035.

So far the output from fusion reactors is not too cheap to meter, just too far in the future, so it's not a solution to our immediate energy problems.

What about nuclear fusion as a solution to our long-term energy problems? First of all, it can only be a partial solution. It is only used for generating electricity, which itself is only 17% of the total energy used in the United Kingdom. Compare this with oil, which accounts for 43% of the energy used in the UK.[37] Nevertheless, electricity from

....

[36]*http://www.fusion.org.uk/culham/jet.html*

[37]*http://www.dti.gov.uk/energy/statistics/source/total/page18424.html DTI analysis of energy by source for 2004. Oil accounts for 47%, gas for 33% and electricity for 17%. This is after adjusting for the oil, coal and gas used for electricity generation.*

nuclear fusion, using fuels which are cheap and abundant, would save the import of coal and gas from other countries and reduce CO_2 emissions – in the future.

Staying with nuclear fission – which is after all available now – an important issue is the time needed to build new nuclear stations. All but one of the existing UK stations will cease all production by 2023 and no new ones are planned, (though the British government's energy review, published in July 2006, strongly suggests that this is soon to change.)[38] Even so, the government itself admits that it takes 10-15 years to complete the construction of a new station.

Coal power stations are also coming to the end of their lives, adding to the need to renew our generating capacity or to replace it with other energy sources. Nuclear stations typically take at least ten years to build, including the time needed for approvals. This might be shortened by applying to build on existing sites or by government action to curtail the inquiry process. Even so, it's unlikely that any new nuclear stations could be ready before 2016.

An alternative is additional gas fired stations: cheap and quick to build and much cleaner than coal-fired stations - always assuming they can get the gas! Once built, however, nuclear has two advantages over gas and coal. First, there are no greenhouse gas emissions from nuclear stations, and no particulates, soot or other gases. Secondly, the supply of fuel for nuclear power stations is largely secure.

SECURITY OF SUPPLY

◊ Although uranium has to be imported, the major producers are Canada and Australia: both stable and friendly countries. Given the small volumes needed, it would be possible to store a year's supply of fuel at a time.

◊ Coal is also imported – some 36 million tonnes per year – so while some stocks can be held, the process is vulnerable to both delivery and price.

....

[38] *http://www.dti.gov.uk/energy/review/index.html*

◊ Gas is increasingly imported – only ten days' stock is currently held in the UK.

◊ The cost of decommissioning nuclear plants, including the long-term storage of waste, cannot be ignored. We already have waste in store, so we already have the commitment to storage whether or not we build new nuclear stations. Figures of £60 billion have been quoted for decommissioning, while failing to mention that this will be spread over 30 years.

In Britain, 3,000 people die on the roads each year, 60 a week, eight or nine every day. 50,000 are injured – some of them permanently disabled. Road transport is cheap and convenient and part of our lives. We accept that it involves risks. We would never give up our cars. Less than 70 people died as a result of the Chernobyl accident. There has been no other nuclear accident with anything approaching this level of fatality, but there are accidents in mines and the oil industry.

The fact is that in ten or fifteen years from now we will need new power stations.

◊ Do we build new coal stations and pollute with CO_2 and other gases?

◊ Do we go for gas and hope that foreign suppliers will continue to supply?

◊ Or do we go for nuclear – perhaps most expensive, most controversial, but cleanest, arguably the safest and with security of supply?

◊ Will renewables fill the gap?

◊ What do we do about supplies for our transport fleet, where coal, gas and nuclear offer no solution at all?

In the next chapter we look at the prospects for alternative energy.

What's the Alternative?
WILL CLIMATE CHANGE YOUR LIFE?

Renewables, rubbish, biofuels and hydrogen

Alternative energy sources take two forms – renewables that use the power of nature directly and biofuels that are derived from plants or organic materials. There are also significant opportunities to recover energy from the waste we dispose of every year.

RENEWABLES

Renewable energy sourced direct from nature can be inconsistent and unreliable. 'Direct from nature' means winds, waves, tides and sunshine. Some of these are closely predictable like tides. Others depend on the weather, the location and the time of day. Sourcing energy direct from nature has problems of intermittence.

Intermittence

Intermittence describes the fact that such energy is not available 24/7, though electricity demand follows a predictable path over the course of the day. Electricity cannot be easily or cheaply stored. Battery technology has improved dramatically over recent years, but it is never likely to be practical or cost effective to store enough power for domestic or industrial use. For example, the battery in your mobile phone may keep it on standby for three or four days but would run an electric kettle for less than two seconds. Not enough for a cup of tea.[39]

....

[39] *That is partly, of course, because the electric kettle is one of the heaviest consumers of power. The energy to run an electric kettle for two seconds would run an energy saving light-bulb for 9 minutes. The power to boil 1 pint of water would light that bulb for nearly 9 hours!*

Storage is not impossible and there are pump storage schemes which are used to supply short-term bursts at peak demand. A pump storage scheme consists of a reservoir and a lake somewhere in the mountains, connected by pipes and turbines. Electricity from conventional power stations is used at times of low demand to pump water from the lake up to the reservoir. When demand peaks, the water is released and pours back down to the lake, through the turbines, generating electricity. A conventional power station takes hours to reach its maximum output. A pump storage scheme can reach maximum output in seconds.

Pump storage is not efficient. There are losses in sending the electricity across the grid to run the pumps, there are losses in pumping, there are losses in the turbines and there are losses in sending the power back across the grid. Pump storage is a very effective way of responding rapidly to peaks in demand, but not a cost-effective way of storing electricity.

If power cannot be stored cost-effectively it has to be used as soon as it is generated. If the supply of renewable energy is fluctuating, because a cloud goes across the sun or the wind drops for a moment, power has to be found from elsewhere or the lights will dim and trains will slow down.

In practice, if renewable energy is being fed into the national grid and accounts for only 6% of total supply as at present, fluctuations can be absorbed. Some industrial customers have interruptible tariffs, which means that the equipment can be turned off remotely by the power supplier when demand exceeds supply. This applies to things like freezers, which can be turned off without harm for short periods; long enough to match demand with supply.

If the proportion of electricity sourced from renewables grows significantly, cutting demand momentarily will not be enough to smooth out the fluctuations in supply. Opponents of renewables argue that the only solution to this problem of intermittency will be to build equivalent capacity of conventional power stations to cover the fluctuations. When the output from wind, waves or solar falls, a gas, coal or nuclear station increases power to make up the shortfall.

"Why," ask the opponents, "why not just build the conventional stations and let them deal with the whole demand?" Trying to make a power station match the fluctuations of renewables is difficult and not an efficient way of operating it.

Others argue that balancing renewables and conventional generation is possible without duplicating the two systems one-for-one, but that renewables can never replace more than a proportion of conventional capacity.

Solar panels

As I write this on the hottest day of July 2006 I wonder about solar panels. I wonder about people who tell me they enjoy sitting in the sun, watching their electricity meters go backwards as they sell solar electricity back to the grid. So how practical is solar energy in Britain, where hot sunny days are the ones we remember because they come so rarely? In the next chapter we consider what we can do at home, but first let's look at some of the options on a commercial scale.

It has been said that if we were to cover only part of the Sahara desert with solar panels we would generate more than enough energy to satisfy the needs of the whole world. Of course there would be substantial costs in manufacturing hundreds of square miles of panels and in getting them to site and installing them. Tens of thousands of miles of pylon lines would be needed, and thousands of miles of undersea cables crossing the oceans. So far nobody is seriously planning to invest.

Most solar projects are on a much smaller scale.

Water heating

The simplest form of solar energy capture is the flat panel collector for water heating. Installed on a roof facing the sun, it can produce hot water for all domestic needs in the summer and, on many days in the winter it will pre-heat the water, reducing the use of electric or gas heating.

Photovoltaic panels

Photovoltaic (PV) panels turn sunlight directly into electricity.
According to the Renewable Energy Association, 'A plane inclined at
about 30 degrees, facing due south ranges from around 900kWh/m2
per year in the North of Scotland to around 1,250kWh/m2 in the
South West of England.' Presumably this means that this amount of
energy will be generated by the panels. The amount that is actually
produced will always depend on the weather. PV panels can be part
of a combined solution as discussed below, but will not supply the
total needs of a household.

PV panels supply electricity directly to the property. If they produce
more than required, then it passes through an output meter into
the grid as power companies have a legal obligation to purchase
electricity supplied in this way. Homes or businesses generating solar
energy are entitled to a Renewable Obligation Certificate for each
1,000kWh of renewable power generated, regardless of whether
they have consumed the energy or sold it back to the grid. These
certificates may be sold back to the power company, to help them
meet their own renewable obligations.[40]

On their own, solar panels cannot supply our total energy needs, but
they can form part of an integrated solution. Examples are discussed
below.

Passive solar design

If you are building a new house, office or factory, passive solar design
principles are important to consider. These involve constructing the
building to take full advantage of sunlight and of heat generated
within the building from lighting, cooking, industrial processes,
computers and people. Solar panels may be part of the solution, but
the main thrust is to construct the building itself as a heat collector
and store. Orienting it towards the south and incorporating internal
walls within a glazed atrium to store the sun's heat, combined
with a carefully-designed ventilation system, can maintain even
temperatures throughout the building day and night. High standards

....
[40]**www.ofgem.gov.uk** *The Office of Gas and Electricity Markets*

of insulation and double-glazing ensure that the minimum energy is wasted. Even in the UK, buildings have been constructed which stay warm at all times of year, relying only on solar heating.

Passive solar design is an important consideration if you are indeed building a new property, though current levels of thermal efficiency demanded by the building regulations are way below what is possible. The government could make a difference by insisting on higher standards. There may be resistance from the construction industry as higher standards would increase their costs, though by far less than the extra cost of heating the building over its lifetime. Heating costs are of course paid by the occupier, not the builder.

Existing structures cannot usually be upgraded to the same levels of efficiency as new build, though insulation can pay for itself and there are still a surprising number of British houses with un-insulated cavity walls.

Wind

Britain has 40% of all the wind available in Europe. This is presumably because we are a group of islands stuck out in the North Atlantic. Although we have all this wind available, Germany is currently the leader in wind power in Europe.

There are plans to change all that. For example, there is a plan to build the biggest wind-farm in Europe on the island of Lewis in the Outer Hebrides. Having lived there for 6 months, I can tell you that it's a bleak expanse of rock and moorland out in the Atlantic off the north-west coast of Scotland and the wind blows nearly all the time. The word 'god-forsaken' may come to mind, but you'd be very wrong. The main town, Stornoway (pop. 6,000), has churches of at least 12 different denominations and they are all full on Sundays; maybe because absolutely everything else is shut!

Lewis Wind Power has applied to install 234 turbines, which will provide 702 MW, about the same as a small power station. By comparison, the Sizewell B nuclear station produces 1,188MW and Drax, the UK's largest coal-fired plant, produces 4,000 MW.

We have already looked at intermittency, which means that wind power, sunlight and tides cannot produce a constant supply of energy at every hour of the day. We can still be sure that the wind will blow on average for a given number of days each year, and the amount of energy that can be harvested is considerable. The wind is free, and it's not controlled by anyone else.

The Lewis wind farm could produce about 1% of the UK's electricity needs at a competitive cost. Of course, 'competitive' depends on how much other fuels cost. The cost of installing the turbines may be more than the cost of other sorts of power plant, but the wind is free.

How close is this project to supplying us with electricity? The application has been submitted but it is likely to be another five years before we can expect construction to begin. There is opposition. This comes from people who are opposed to wind-farms in general, from people who believe they will be a danger to birds and from people who live on the island.

Their first concern is that the 234 turbines will be intrusive. Certainly at 140m high with rotors describing a 100m diameter they will be very visible. Although some of the connecting cables will be laid underground, there will be a 35 mile run of 141 pylons, each of which will be 27m high. To construct and maintain the turbines will require the building of 104 miles of road. There will also be nine substations, a control building and various other structures. All this will be built on the island, but since the output of the farm will be far more than is required by the local population, it is planned to send it to Glasgow and Edinburgh. To do this, the first step is to install an undersea line to carry the current ashore.

At present, Lewis gets electricity from the mainland via a submarine cable, but this does not have sufficient capacity to carry the output of the wind-farm. Once the new submarine cable is in place, the next stage is to build a pylon route through the Highlands. It will stretch for some 50 miles from Ullapool on the west coast to link to the grid at Beauly near Inverness. Reportedly the pylons will be 50m high (nearly twice the size of those proposed for the island), and needless to say, there is vociferous opposition to building them.[41]

....

[41]*http://www.hbp.org.uk/* Highlands before Pylons

Until this is resolved there will be no commitment to start any sort of construction on Lewis. It will probably be ten years before we can reasonably expect the electricity to flow from the Lewis Wind Farm, and that's if all goes well. That's four or five years after some of our existing power stations will have been decommissioned.

In the next chapter we look at the feasibility of installing a wind turbine at home.

Tides

Tidal power is an attractive form of renewable energy because it is totally predictable. Tides rise and fall twice a day, every day – though not at exactly the same time every day.

There are two main ways of harvesting the energy from these masses of moving water. The first is the tidal barrage. Built across an estuary, the barrage houses turbines. These are turned as the tide comes in and the water passes through the barrage. As the tide falls and the water flows out they are turned again.

Electricité de France constructed a barrage across the Rance estuary in the 1960s, which continues to produce 240MW from its 10 turbines.[42] By comparison, a small coal or nuclear power station produces around 750MW.

A tidal barrage needs a rise and fall of at least 5m: at Rance the daily movement is 8m. Britain has some of the highest tides in the world and a barrage across the Severn Estuary has been considered for many years. Arguments against such a barrage include the obstruction it would cause to shipping, depending on exactly where it was built; damage to ecosystems and wildlife and silting up of parts of the river. All these issues have been addressed at Rance, and while there has been some silting in the estuary, the overall environmental impact has been small. A Severn barrage would be very much bigger, and much more expensive, than the one at Rance, which may be a reason why barrages have a low priority in the UK.

. . . .

[42] **http://www.edf.fr/html/en/decouvertes/voyage/usine/retour-usine.html** *Rance Barrage home page.*

The alternative to barrages is to exploit deep-water tidal flows.[43] Some of the world's strongest tidal flows are found off the shores of the United Kingdom, principally around the Channel Islands and the Pentland Firth in Scotland.

There are several types of turbine, but those under active consideration at present are much like wind turbines, mounted on the sea bed as much as 40m below the surface. Cost has been one of the main factors holding back the development of tidal power. While electricity from other sources has been relatively cheap, there has been no point in examining alternatives which can never be competitive. Now that prices of oil and gas are on the rise and there are concerns that more and more of our energy is being imported, tidal power begins to look more interesting.

Cost will still be a consideration. Installing and maintaining turbines at the bottom of the sea in areas where the tides are particularly strong will always be difficult, dangerous and expensive. The turbines need a grid connection as well – an undersea cable and pylons once onshore. A study commissioned by the Carbon Trust estimates that despite the abundance of tidal power in the UK, output would be around 18tWh/year, or only 5% of UK demand.

Tidal power is another solution with potential, but nowhere near an answer on its own.

ENERGY FROM WASTE

The UK generates some 30m tonnes of rubbish each year and 84% goes to holes in the ground – landfill. This is one of the highest proportions in Europe and is probably influenced by the fact that landfill tax is significantly lower in the UK than in other countries. Rubbish can be a source of pollution, but also a source of energy. While we are importing coal that has been dug up in some foreign land, at the same time we are burying rubbish with a usable energy content. The energy in British rubbish is equivalent to 35m tonnes of coal.[44] In any case, landfill sites are becoming more difficult to

....

[43]*http://www.carbontrust.co.uk/technology/technologyaccelerator/tidal_stream2.htm*
Carbon Trust
[44]*Dr Frank Hardwick, Environment Agency in a presentation to Envirenergy 2006, September 2006, Leeds.*

The industrial process produces a consistent product which will not clog up the system in cold weather. Have you ever seen what happens to cooking oil in the fridge or a cold kitchen? It goes all white and waxy. This is what could happen to your car after a cold night unless you are using commercially-produced bio-diesel. You'll have a job to start it before the spring.

Commercially-produced bio-diesel has a lot to recommend it. It's a clean fuel that can be made from organic waste, which would otherwise have to be disposed of – at a cost. Although it is currently heavily diluted with mineral diesel before sale, it could be used successfully on its own in most diesel vehicles. However, it can only ever be a small part of our energy supply. The DETR (Department of Energy Transport and Regions) estimates that there are around 100,000 tonnes of suitable waste, which could yield some 110 million litres of fuel. That's about 0.6% of the 20 billion litres used annually in the UK.

Production could be increased if we grew crops of oilseed rape for the purpose, but that would take energy for the fertilisers, the planting, the harvesting, the transport, the extraction of the oil and disposal of the waste. Even if we could increase bio-diesel output ten-fold (assuming we could find all the land needed), that would yield only a tiny proportion of total demand, and all those costs would have to be built into the price at the pump.

Bio-ethanol is a substitute fuel for petrol engines, distilled from organic matter such as sugar cane, wheat or maize. Brazil has an extensive bio-ethanol programme based on sugar, while the United States is supporting farmers to produce corn [maize] for ethanol production.

The European Commission announced in June 2006 that it would spend €130 million to purchase 510 million litres of surplus wine and turn that into bio-ethanol. Wine is only about 12% alcohol, which means that the remaining 88% (water) has to be extracted and thrown away. Given that this wine is already a subsidised product; the production of bio-ethanol from wine cannot be remotely cost-effective. Of course the costs may not be truly reflected at the pump – they will be paid by the European taxpayer.

find and there is more and more pressure to recycle the materials in rubbish rather than just dump them. Energy from rubbish must be part of an overall recycling and waste management policy.

Anaerobic digestion

Some of the things we throw away are biodegradable, which means they rot down by natural processes. They do this in two ways: aerobic and anaerobic digestion.

Aerobic digestion is what happens when organic material is left open to the air and rots. This is what goes on a garden compost heap, and as the digestion proceeds it releases CO_2 and methane. (Spreading the compost on the garden releases more, so perhaps it's not as green as you thought!) Local councils compost substantial quantities of organic material, but the value of the material produced is generally far less than the cost of production. That is probably why they encourage people to compost at home by offering free compost bins.

Anaerobic digestion occurs when substances break down without exposure to air. Much of the waste sent to landfill is organic, and decomposes and rots in this way. The process generates significant quantities of methane over time, which may leak away into the atmosphere. Methane, you may remember, is a greenhouse gas 23 times as strong as CO_2. Wherever possible the methane is collected and may be flared off. At least burning it produces CO_2, less harmful than methane. In fact substantial volumes of methane are captured and currently yield some 650MW of energy. The potential is believed to be as much as 2-3 terawatts (Tw), which is a significant proportion of the 65 – 70Tw of electricity generated in the UK.

Recycling and incineration

Relying on rubbish to rot in landfill sites in order to produce energy via landfill gas is a crude and inefficient process. Apart from anything else, all the material that does not rot, which may include metals and plastics, is simply abandoned.

Recycling, based on sorting the waste by type, can yield materials which can be re-used, materials which can be burnt for energy and materials which can be digested for gas. Some materials can be liquefied and used as fuel. Spontaneous anaerobic digestion, which happens by itself in a landfill, can take weeks. Digestion in special vessels can take days, but new technology is under development to achieve this in hours, dramatically expanding plant capacity.

The gas can be used to run an engine to drive an electrical generator, it can be used as road vehicle fuel or it can be put directly into the gas main. Where the gas runs an engine, the output is heat as well as electricity.

Burning waste raises doubts, emotions and opposition. Many people are implacably opposed to municipal incinerators, which they see as polluting and poisonous. There is no doubt that imperfectly controlled combustion has caused dioxins and other materials to be emitted from incinerators in the past, but this has been recognised and new designs stringently restrict the gases and particulates emitted.

Proponents will tell you that you would get more dioxins from a patio barbecue than from the top of a modern incinerator chimney. Opponents argue that everything should be recycled, but there are some things that cannot be recycled, and the best way to recover energy from them and to minimise the residue sent to landfill is to incinerate them with care.

Public policy

Waste disposal in the UK is currently in the hands of local authorities with no overall government policy or control. European directives demand a consistent approach to waste, but the government has been slow to implement them.

A classic example is the repeatedly postponed introduction of the Waste Electrical and Electronic Equipment Disposal directive, which has yet to come into force. For some years the industry has been urging the government to give a lead, in the face of a shortage of landfill sites, a very low level of recycling compared with other European countries, wasted energy and public opposition based on ignorance of the issues and opportunities.

BIOFUELS

The second main group of alternative energy sources includes things like bio-diesel, bio-ethanol and biomass. When these are burned they release carbon dioxide, but if the plants are replaced they take up carbon dioxide as they grow so, on the face of it, there's no net increase in CO_2. There's no intermittence problem as any of these can produce continuous energy 24/7. They can be stored and used as and when required.

Bio-diesel, bio-ethanol and used cooking oil

If you have a diesel car there's nothing to stop you filling your tank with used cooking oil and driving it away. Theoretically. In fact, you should declare what you're doing to the tax authorities and pay a special rate of road fuel duty on it. This annoys a lot of people who have found that cooking oil works perfectly well and costs next to nothing. Used cooking oil is effectively bio-diesel, and surely it's environmentally friendly? It produces CO_2 when burnt, but it comes from plants – maize, sunflowers and so on – so when more plants are grown to make more oil, the CO_2 is absorbed again.

Argent Energy[45] operates the UK's first large-scale bio-diesel plant. It takes used cooking oil and other organic waste and puts it through a series of chemical processes and filtrations to create the final product. It is mixed with ordinary mineral diesel in the proportion of 5% bio to 95% diesel by the oil companies and sold as normal road fuel.

Why is all this processing necessary when people can run their cars on straight used cooking oil? Typically they filter it before use, but the problem with cooking oil is variability of quality. Also, it's not just CO_2 that's produced when cooking oil is burnt in an engine. The industrial process removes the sulphur and the other impurities which would otherwise produce polluting gases. It removes the water which is often found in waste oil, which can damage the fuel tank, engine, exhaust system and catalytic converter. (No certificate of roadworthiness or MoT if that's not working!)

....

[45]http://www.argentenergy.com

Biomass

Biomass is plant material which is used to generate electricity. It may be burnt to produce heat to raise steam to drive a turbine and generator. It may be gasified for use in an internal combustion engine or turned into oil or charcoal. For example, one type of plant that is widely used, coppice willow, can grow on a wide range of soils, including contaminated land, as long as there is plenty of water. The top growth is harvested every 3-5 years and the plants will re-grow five or six times.[46] Elephant grass or miscanthus is another rapid-growing plant with a high energy content. Like willow, it can be gasified or simply dried and burnt.

ADAS UK Ltd, an environmental and agricultural consultancy,[47] has estimated that 20% of UK agricultural land area could produce crops yielding enough energy for 8 million homes – one-third of all UK homes. And they claim that all of this would be carbon neutral, since the growing plants absorb as much CO_2 as is released when they are used for energy. However, this is only part of the story.

Inputs, outputs and total production

Like oilseed rape for bio-diesel, the crops for bio-ethanol and biomass need to be planted, fertilised, harvested, transported and processed. All this takes energy, so the net energy output from biofuels is reduced by all the energy consumed in the production cycle. In fact, there is also wide scepticism of biofuels' 'carbon-neutral' credentials and the other benefits they are claimed to bring.[48]

Brazil uses sugar cane to produce 50% of the world's ethanol and all petrol sold in the country contains 20% bio-ethanol. Brazil needs only 3% of its agricultural land to produce 10% of its fuel requirements. Unfortunately for the rest of the world, this is a special case. Brazil has an ideal climate for growing sugar cane, it has a low population density and Brazilians do not drive nearly as far as people

....

[46] *The Scottish Agricultural College* **http://www1.sac.ac.uk/envsci/External/Willow-Power/Willow_s.pdf**

[47] **http://www.adas.co.uk/bluesky35/energy.html**

[48] *New Scientist 23rd September 2006*

in other countries. For these reasons the United States would need to use 30% of its agricultural land for enough biofuel to meet 10% of its needs: for the same proportion, Europe would have to use 72% of its agricultural land. In fact, the EU has the stated objective of including 5.75% of biofuel in all fuels by 2010. To do this it will need to import vegetable oils as feedstock as there is not enough suitable land available to grow biofuel crops.

The US, as we have seen, is encouraging farmers in the Midwest to grow corn for ethanol production and is paying subsidies. By 2007 20% of the US corn crop will be converted into ethanol. Some researchers believe that this is a fruitless exercise and that the corn/ethanol production cycle – ploughing, fertilising, processing, transporting - will use 30% **more** energy than it actually produces.[49] Others see a net benefit, but nobody claims that the process is anything like 100% carbon-neutral or energy-neutral.[50]

There are wider issues; principally the question of whether we should use food crops for fuel when there are famines and shortages elsewhere in the world. If the US converts corn to ethanol there will be less corn on the world market and higher prices for those who cannot grow their own. If climate change causes drought and famine, the numbers of those without food can only grow.

Sugar cane is a more efficient crop for conversion into ethanol but it is a crop which demands substantial quantities of water to grow. In India this has led to lowered water tables and competition with food crops. In Brazil expanding sugar cane for ethanol has meant that farmers cut down the rain forest for land to grow other crops.

Palm oil is even better for making biofuels, but expanding palm oil cultivation in Malaysia has brought extensive rain forest destruction. Cutting down the rain forest destroys the habitat of rare species. It reduces the most important carbon sink on the planet: the plants that keep greenhouses gases in check by absorbing CO_2. The rain forest absorbs the rains. Without the forest, the monsoon can wash away

· ·

[49] *Professor David Pimentel, Cornell University*
http://www.news.cornell.edu/stories/July05/ethanol.toocostly.ssl.html
[50] *Alexander Farrell and others, University of California*
http://www.sciencemag.org/cgi/content/short/311/5760/506

the fertile surface soil for ever. As trees are cut down and burnt or left to rot they release CO_2 to the atmosphere. In recent years forest fires set by people clearing land for oil palms have run wild. For hundreds of square miles the sun has been obscured by smoke clouds big enough to be seen from space. Palm oil production may be profitable in money terms, but producing it for biofuel causes far more damage than the fossil fuel it replaces.

From a security point of view, growing biofuel crops like oilseed rape in Britain has the advantage that they are a secure source of supply independent of imports, of foreign states and of foreign exchange rates. However, the government's target is to source 5% of all road fuel from bio-fuels. If we grow enough crops for this and also grow enough biomass to power those 8 million homes mentioned above, we are left with less than 60% of land for growing food.

Of course, not all UK land is economic for growing food and not all our land is used at present for growing food. Indeed, over 40% of the food, feed and drink consumed in the UK is imported.[51]

◊ Some of these are products which cannot be produced in the UK.

◊ Some of them are imported so that we can enjoy them year-round, regardless of seasons.

◊ Some of them come from third-world economies where the cost of production is so low that the goods are still cheaper after shipping them from the other side of the world.

All of them must be transported here, using aircraft, ships, road and rail. All of this uses energy.

So if we grow crops to use for biofuels, and we stop growing food and have to import it, are we really making any savings?

It is simplistic to say that we should just stop importing food, grow it here and save the energy needed to transport it. We cannot replace

....

[51]By value, 2002. See DEFRA
http://statistics.defra.gov.uk/esg/reports/afq/afqbriefsup_dec.pdf

all imported food with home production. We may be able to grow enough crops to provide 5% of our road fuel requirements, but that still leaves 95% dependent on the North Sea and foreign suppliers.

But is there nothing else that can solve the problem? 'The Stone Age did not end for lack of stones'[52] – it was overtaken by something better. Is there now a wonder fuel poised to take over from oil?

HYDROGEN

At first sight, hydrogen is the future. It's clean. When you burn it, the only by-product is pure water. It's abundant. It's one of the most common elements in nature. It's easy to use. With minor modifications, a conventional internal combustion engine can use it. Fuel cells – silent, and with no moving parts – turn hydrogen to electricity, leaving nothing but water.

Too good to be true? Hydrogen is abundant, but rarely found on its own. In other words it exists in chemical compounds, the most common of which is water. Hydrogen is a store of energy, not a readily available source of energy. Hydrogen can be produced by stripping it out of natural gas or coal. This process takes energy and leaves by-products. In both cases it releases CO_2, and at present this is generally vented into the atmosphere.

If coal is the source, all the other by-products, the sulphur and tar, have to be disposed of as well. The simplest way of producing hydrogen is to electrolyse water. Passing a current through water splits it into hydrogen and oxygen. The oxygen can blow away into the atmosphere. It's not toxic, it's not a green house gas and it will get used up again when the hydrogen is burnt.

Unfortunately, no chemical process is perfect, so it may not be a surprise to discover that the energy stored in the hydrogen is much less than the energy needed to release it from the water. The electrolysis process releases a lot of heat as a wasted by-product.

....

[52]*Said by Sheik Yamani, famous as Saudi Oil Minister at the time of the first oil crisis in the 1970s when the OPEC nations reduced supply and raised prices.*

find and there is more and more pressure to recycle the materials in rubbish rather than just dump them. Energy from rubbish must be part of an overall recycling and waste management policy.

Anaerobic digestion

Some of the things we throw away are biodegradable, which means they rot down by natural processes. They do this in two ways: aerobic and anaerobic digestion.

Aerobic digestion is what happens when organic material is left open to the air and rots. This is what goes on a garden compost heap, and as the digestion proceeds it releases CO_2 and methane. (Spreading the compost on the garden releases more, so perhaps it's not as green as you thought!) Local councils compost substantial quantities of organic material, but the value of the material produced is generally far less than the cost of production. That is probably why they encourage people to compost at home by offering free compost bins.

Anaerobic digestion occurs when substances break down without exposure to air. Much of the waste sent to landfill is organic, and decomposes and rots in this way. The process generates significant quantities of methane over time, which may leak away into the atmosphere. Methane, you may remember, is a greenhouse gas 23 times as strong as CO_2. Wherever possible the methane is collected and may be flared off. At least burning it produces CO_2, less harmful than methane. In fact substantial volumes of methane are captured and currently yield some 650MW of energy. The potential is believed to be as much as 2-3 terawatts (Tw), which is a significant proportion of the 65 – 70Tw of electricity generated in the UK.

Recycling and incineration

Relying on rubbish to rot in landfill sites in order to produce energy via landfill gas is a crude and inefficient process. Apart from anything else, all the material that does not rot, which may include metals and plastics, is simply abandoned.

Recycling, based on sorting the waste by type, can yield materials which can be re-used, materials which can be burnt for energy and materials which can be digested for gas. Some materials can be liquefied and used as fuel. Spontaneous anaerobic digestion, which happens by itself in a landfill, can take weeks. Digestion in special vessels can take days, but new technology is under development to achieve this in hours, dramatically expanding plant capacity.

The gas can be used to run an engine to drive an electrical generator, it can be used as road vehicle fuel or it can be put directly into the gas main. Where the gas runs an engine, the output is heat as well as electricity.

Burning waste raises doubts, emotions and opposition. Many people are implacably opposed to municipal incinerators, which they see as polluting and poisonous. There is no doubt that imperfectly controlled combustion has caused dioxins and other materials to be emitted from incinerators in the past, but this has been recognised and new designs stringently restrict the gases and particulates emitted.

Proponents will tell you that you would get more dioxins from a patio barbecue than from the top of a modern incinerator chimney. Opponents argue that everything should be recycled, but there are some things that cannot be recycled, and the best way to recover energy from them and to minimise the residue sent to landfill is to incinerate them with care.

Public policy

Waste disposal in the UK is currently in the hands of local authorities with no overall government policy or control. European directives demand a consistent approach to waste, but the government has been slow to implement them.

A classic example is the repeatedly postponed introduction of the Waste Electrical and Electronic Equipment Disposal directive, which has yet to come into force. For some years the industry has been urging the government to give a lead, in the face of a shortage of landfill sites, a very low level of recycling compared with other European countries, wasted energy and public opposition based on ignorance of the issues and opportunities.

BIOFUELS

The second main group of alternative energy sources includes things like bio-diesel, bio-ethanol and biomass. When these are burned they release carbon dioxide, but if the plants are replaced they take up carbon dioxide as they grow so, on the face of it, there's no net increase in CO_2. There's no intermittence problem as any of these can produce continuous energy 24/7. They can be stored and used as and when required.

Bio-diesel, bio-ethanol and used cooking oil

If you have a diesel car there's nothing to stop you filling your tank with used cooking oil and driving it away. Theoretically. In fact, you should declare what you're doing to the tax authorities and pay a special rate of road fuel duty on it. This annoys a lot of people who have found that cooking oil works perfectly well and costs next to nothing. Used cooking oil is effectively bio-diesel, and surely it's environmentally friendly? It produces CO_2 when burnt, but it comes from plants – maize, sunflowers and so on – so when more plants are grown to make more oil, the CO_2 is absorbed again.

Argent Energy[45] operates the UK's first large-scale bio-diesel plant. It takes used cooking oil and other organic waste and puts it through a series of chemical processes and filtrations to create the final product. It is mixed with ordinary mineral diesel in the proportion of 5% bio to 95% diesel by the oil companies and sold as normal road fuel.

Why is all this processing necessary when people can run their cars on straight used cooking oil? Typically they filter it before use, but the problem with cooking oil is variability of quality. Also, it's not just CO_2 that's produced when cooking oil is burnt in an engine. The industrial process removes the sulphur and the other impurities which would otherwise produce polluting gases. It removes the water which is often found in waste oil, which can damage the fuel tank, engine, exhaust system and catalytic converter. (No certificate of roadworthiness or MoT if that's not working!)

....

[45]*http://www.argentenergy.com*

The industrial process produces a consistent product which will not clog up the system in cold weather. Have you ever seen what happens to cooking oil in the fridge or a cold kitchen? It goes all white and waxy. This is what could happen to your car after a cold night unless you are using commercially-produced bio-diesel. You'll have a job to start it before the spring.

Commercially-produced bio-diesel has a lot to recommend it. It's a clean fuel that can be made from organic waste, which would otherwise have to be disposed of – at a cost. Although it is currently heavily diluted with mineral diesel before sale, it could be used successfully on its own in most diesel vehicles. However, it can only ever be a small part of our energy supply. The DETR (Department of Energy Transport and Regions) estimates that there are around 100,000 tonnes of suitable waste, which could yield some 110 million litres of fuel. That's about 0.6% of the 20 billion litres used annually in the UK.

Production could be increased if we grew crops of oilseed rape for the purpose, but that would take energy for the fertilisers, the planting, the harvesting, the transport, the extraction of the oil and disposal of the waste. Even if we could increase bio-diesel output ten-fold (assuming we could find all the land needed), that would yield only a tiny proportion of total demand, and all those costs would have to be built into the price at the pump.

Bio-ethanol is a substitute fuel for petrol engines, distilled from organic matter such as sugar cane, wheat or maize. Brazil has an extensive bio-ethanol programme based on sugar, while the United States is supporting farmers to produce corn [maize] for ethanol production.

The European Commission announced in June 2006 that it would spend €130 million to purchase 510 million litres of surplus wine and turn that into bio-ethanol. Wine is only about 12% alcohol, which means that the remaining 88% (water) has to be extracted and thrown away. Given that this wine is already a subsidised product; the production of bio-ethanol from wine cannot be remotely cost-effective. Of course the costs may not be truly reflected at the pump – they will be paid by the European taxpayer.

Biomass

Biomass is plant material which is used to generate electricity. It may be burnt to produce heat to raise steam to drive a turbine and generator. It may be gasified for use in an internal combustion engine or turned into oil or charcoal. For example, one type of plant that is widely used, coppice willow, can grow on a wide range of soils, including contaminated land, as long as there is plenty of water. The top growth is harvested every 3-5 years and the plants will re-grow five or six times.[46] Elephant grass or miscanthus is another rapid-growing plant with a high energy content. Like willow, it can be gasified or simply dried and burnt.

ADAS UK Ltd, an environmental and agricultural consultancy,[47] has estimated that 20% of UK agricultural land area could produce crops yielding enough energy for 8 million homes – one-third of all UK homes. And they claim that all of this would be carbon neutral, since the growing plants absorb as much CO_2 as is released when they are used for energy. However, this is only part of the story.

Inputs, outputs and total production

Like oilseed rape for bio-diesel, the crops for bio-ethanol and biomass need to be planted, fertilised, harvested, transported and processed. All this takes energy, so the net energy output from biofuels is reduced by all the energy consumed in the production cycle. In fact, there is also wide scepticism of biofuels' 'carbon-neutral' credentials and the other benefits they are claimed to bring.[48]

Brazil uses sugar cane to produce 50% of the world's ethanol and all petrol sold in the country contains 20% bio-ethanol. Brazil needs only 3% of its agricultural land to produce 10% of its fuel requirements. Unfortunately for the rest of the world, this is a special case. Brazil has an ideal climate for growing sugar cane, it has a low population density and Brazilians do not drive nearly as far as people

....

[46]The Scottish Agricultural College *http://www1.sac.ac.uk/envsci/External/Willow-Power/Willow_s.pdf*

[47]*http://www.adas.co.uk/bluesky35/energy.html*

[48]*New Scientist 23rd September 2006*

in other countries. For these reasons the United States would need to use 30% of its agricultural land for enough biofuel to meet 10% of its needs: for the same proportion, Europe would have to use 72% of its agricultural land. In fact, the EU has the stated objective of including 5.75% of biofuel in all fuels by 2010. To do this it will need to import vegetable oils as feedstock as there is not enough suitable land available to grow biofuel crops.

The US, as we have seen, is encouraging farmers in the Midwest to grow corn for ethanol production and is paying subsidies. By 2007 20% of the US corn crop will be converted into ethanol. Some researchers believe that this is a fruitless exercise and that the corn/ethanol production cycle – ploughing, fertilising, processing, transporting - will use 30% **more** energy than it actually produces.[49] Others see a net benefit, but nobody claims that the process is anything like 100% carbon-neutral or energy-neutral.[50]

There are wider issues; principally the question of whether we should use food crops for fuel when there are famines and shortages elsewhere in the world. If the US converts corn to ethanol there will be less corn on the world market and higher prices for those who cannot grow their own. If climate change causes drought and famine, the numbers of those without food can only grow.

Sugar cane is a more efficient crop for conversion into ethanol but it is a crop which demands substantial quantities of water to grow. In India this has led to lowered water tables and competition with food crops. In Brazil expanding sugar cane for ethanol has meant that farmers cut down the rain forest for land to grow other crops.

Palm oil is even better for making biofuels, but expanding palm oil cultivation in Malaysia has brought extensive rain forest destruction. Cutting down the rain forest destroys the habitat of rare species. It reduces the most important carbon sink on the planet: the plants that keep greenhouses gases in check by absorbing CO_2. The rain forest absorbs the rains. Without the forest, the monsoon can wash away

· ·

[49] *Professor David Pimentel, Cornell University*
http://www.news.cornell.edu/stories/July05/ethanol.toocostly.ssl.html
[50] *Alexander Farrell and others, University of California*
http://www.sciencemag.org/cgi/content/short/311/5760/506

the fertile surface soil for ever. As trees are cut down and burnt or left to rot they release CO_2 to the atmosphere. In recent years forest fires set by people clearing land for oil palms have run wild. For hundreds of square miles the sun has been obscured by smoke clouds big enough to be seen from space. Palm oil production may be profitable in money terms, but producing it for biofuel causes far more damage than the fossil fuel it replaces.

From a security point of view, growing biofuel crops like oilseed rape in Britain has the advantage that they are a secure source of supply independent of imports, of foreign states and of foreign exchange rates. However, the government's target is to source 5% of all road fuel from bio-fuels. If we grow enough crops for this and also grow enough biomass to power those 8 million homes mentioned above, we are left with less than 60% of land for growing food.

Of course, not all UK land is economic for growing food and not all our land is used at present for growing food. Indeed, over 40% of the food, feed and drink consumed in the UK is imported.[51]

◊ Some of these are products which cannot be produced in the UK.

◊ Some of them are imported so that we can enjoy them year-round, regardless of seasons.

◊ Some of them come from third-world economies where the cost of production is so low that the goods are still cheaper after shipping them from the other side of the world.

All of them must be transported here, using aircraft, ships, road and rail. All of this uses energy.

So if we grow crops to use for biofuels, and we stop growing food and have to import it, are we really making any savings?

It is simplistic to say that we should just stop importing food, grow it here and save the energy needed to transport it. We cannot replace

• • • • • • • • • • • • • • • • • • • •

[51]*By value, 2002. See DEFRA*
http://statistics.defra.gov.uk/esg/reports/afq/afqbriefsup_dec.pdf

all imported food with home production. We may be able to grow enough crops to provide 5% of our road fuel requirements, but that still leaves 95% dependent on the North Sea and foreign suppliers.

But is there nothing else that can solve the problem? 'The Stone Age did not end for lack of stones'[52] – it was overtaken by something better. Is there now a wonder fuel poised to take over from oil?

HYDROGEN

At first sight, hydrogen is the future. It's clean. When you burn it, the only by-product is pure water. It's abundant. It's one of the most common elements in nature. It's easy to use. With minor modifications, a conventional internal combustion engine can use it. Fuel cells – silent, and with no moving parts – turn hydrogen to electricity, leaving nothing but water.

Too good to be true? Hydrogen is abundant, but rarely found on its own. In other words it exists in chemical compounds, the most common of which is water. Hydrogen is a store of energy, not a readily available source of energy. Hydrogen can be produced by stripping it out of natural gas or coal. This process takes energy and leaves by-products. In both cases it releases CO_2, and at present this is generally vented into the atmosphere.

If coal is the source, all the other by-products, the sulphur and tar, have to be disposed of as well. The simplest way of producing hydrogen is to electrolyse water. Passing a current through water splits it into hydrogen and oxygen. The oxygen can blow away into the atmosphere. It's not toxic, it's not a green house gas and it will get used up again when the hydrogen is burnt.

Unfortunately, no chemical process is perfect, so it may not be a surprise to discover that the energy stored in the hydrogen is much less than the energy needed to release it from the water. The electrolysis process releases a lot of heat as a wasted by-product.

··························

[52]Said by Sheik Yamani, famous as Saudi Oil Minister at the time of the first oil crisis in the 1970s when the OPEC nations reduced supply and raised prices.

The electricity required has to be generated from something and conventional coal power stations are about 35% efficient. Gas and nuclear are somewhat better, but power is still lost in the grid as the power is transported to the hydrogen generator.

Then there's the issue of storage. At atmospheric pressure, a volume of hydrogen is 3,000 times[53] the size of a volume of petrol containing equivalent energy. To compress it down to something like the size of a car's petrol tank takes more energy, and the special tanks required to take this pressure (and remain safe in an accident!) need to be very strong. A hollow tank is not the best vessel for this purpose.

Hydrogen is the smallest chemical element and its atoms can seep through almost anything, including some metals. Scientists are currently looking at a sort of metal sponge to absorb and store hydrogen, though they admit it will take some years to complete development. I wonder how quickly the hydrogen will come out of the 'sponge' when I put my foot down. How quickly will I be able to refill it?

And talking of refilling – where do I go when the gauge on my hydro-car is pushing empty? At present there are no public filling stations where you can go to fuel your hydrogen-powered vehicle. There are hydrogen supplies in some public transport garages, but they are very rare and strictly experimental. Transport for London is taking part in an international trial.[54]

Are the oil companies going to install facilities at their petrol filling stations? When they could be selling petrol instead? And supposing they do put these special pumps in, how are they going to supply the stations? Even at high pressure, the tankers are going to carry less energy than if they were carrying petrol, so the cost of getting hydrogen to the point of sale from the point of production is likely to be greater.

This raises the question of the point of production: where are we going to produce our hydrogen? Maybe we should produce it at the

....

[53]*BBC News Oct 2004*
[54]*http://www.tfl.gov.uk/buses/fuel-cell-buses.asp*

point of sale: a gas generator at every filling station. All that's needed is a supply of water and a supply of electricity; a holding vessel and some compressors. No need for tankers or for carrying hydrogen around on our roads.

However, we still need to be sure that it's safe. We're going to have to have carefully-trained operators at every site. We'll need to invest in substantial equipment at every filling station. More energy has to be put in to release the hydrogen and then to compress it; more energy goes in than we can expect to get out. More costs. The only person who will pay for all this is the consumer, through the price at the pump. Hydrogen may be an alternative fuel, but will it be cheaper?

Hydrogen can leak, and if it leaks it's odourless, so you won't smell it. Another strange thing about hydrogen is that it burns with an invisible flame, so if it leaks and it catches fire you won't notice it. And if that leads to an explosion, you won't notice that either. Not before it's too late, anyway – and then it'll be too late.

On balance, I cannot see hydrogen being a practical solution to our need for portable energy, to run our cars, buses and other means of transport. However, it may be a useful back-up to renewables and provide static energy for the National Grid.

As discussed above, one thing that renewables all have in common is intermittency, which means that sometimes the sun doesn't shine on the PV panels, sometimes the wind drops and the windmills fall silent and twice a day it's slack tide and the tidal turbines are still. On the other hand, sometimes these natural resources provide more energy than is needed, and at that time they could be used to produce hydrogen, which in turn could turn generators or feed fuel cells when the energy from renewables falls off. This is an attractive way of storing energy which could be fed back into the grid. Maybe it could match pump storage as a means for meeting surges in demand. See the Tees Valley project described in the next chapter.

Some people have taken this a stage further.[55] They see the average consumer generating electricity at home from wind turbines and

....

[55]See the two sides of the argument in Half Gone by Jeremy Leggett, Portobello Books 2005 and Beyond Oil by Kenneth S. Deffeyes, Hill and Wang 2005

solar panels and turning this into hydrogen. The hydrogen goes into their car. They drive to work and park in specially-designed bays where their car becomes a generator. It uses the hydrogen to produce electricity and sells it back to the power grid. It's an attractive solution, but is it practical? Would you want your neighbours to be generating and storing hydrogen - a substance as dangerous as petrol? If you wanted to make an unexpected journey after work, would there be enough hydrogen left in your tank? How many people could actually attach their cars to a source of hydrogen in their houses? Many people live in flats, many in terraces with only on-street parking.

There's a future for hydrogen, but like many other energy sources, as a solution to our energy problems it's not enough on its own. And it seems to be still some way in the future.

PRACTICAL ALTERNATIVES

None of these ideas – renewables, biofuels or hydrogen – seems to be the total answer to energy shortages. However, a number of projects have been successful by using some of these ideas in combination and closely managing energy use for efficiency. Others are in the planning stage. The next chapter looks at some of them in detail.

Putting it into Practice 6
WILL CLIMATE CHANGE YOUR LIFE?

Case studies

You can't open a newspaper without realising that the government wants us to cut CO_2 emissions. They have increased the taxes on large cars, established initiatives and organisations, including the Carbon Trust and the Energy Saving Trust, and even started talking about taxing aviation fuel, so bringing an end to cheap flights. It will be a brave government that actually does that.

Some organisations have been talking about CO_2 for decades and already have long-established schemes using renewables and more efficient ways of using energy to cut costs and reduce emissions.

One of the most popular approaches is Combined Heat and Power (CHP). Generating electricity by any method involves wasted heat, which is why conventional power stations have limited efficiency. You see those great cooling towers in many articles about pollution, but in fact the clouds coming out of them are nothing but steam. There's no pollution, just a lot of wasted energy. The cooling towers are there to recover the water from the exhaust steam from the turbines, so that it can be pumped back to the boilers. As the steam condenses it gives up heat which is lost to the atmosphere. This is part of the reason why the best conventional power stations are still less than 50% efficient.

A CHP system often uses an internal combustion engine burning gas from the normal supply to run a generator, and like any other engine it creates heat. This heat is transferred to a water jacket, and this hot water is circulated to central heating radiators and heat exchangers. Energy from the CHP system can also be used to provide chilled water for air conditioning.

District heating was always very popular in the old Soviet Union, though not necessarily used to its best advantage. Many systems still exist and have been modernised. Instead of having a boiler in every house, there is a heating plant in every district or apartment block. Heavily insulated hot water mains are laid in every street and piped to every property where each dwelling has a heat exchanger instead of its own boiler. This reduces the heat from the temperature of the main and feeds it to the domestic central heating system as required. The heat is metered like any other utility. A single central boiler is more efficient and less polluting than a boiler in every house. There is only one boiler to be maintained, and the individual heat exchangers are cheaper and more reliable than individual boilers.

Critics argue that if the one central boiler fails then everyone gets cold. In practice there are back-up systems to minimise the risk. Several of the schemes described below use CHP to supply a district heating system.

WOKING BOROUGH COUNCIL[56]

Woking's investment in alternative energy includes CHP, PV solar panels, and a fuel cell. The Council has been working on this since 1990 and approached sustainable and renewable energy projects in four key ways:

◊ Initial energy efficiency policy

◊ Formation of energy and environmental services companies

◊ Cost neutrality to Council Tax taxpayer

◊ Innovative funding opportunities

....

[56]*Woking's Sustainable and Renewable Energy Installations –*
http://www.woking.gov.uk

These have created a successful energy-efficient and cost-efficient provision of services to the residents in the borough.

Combined heat and power (CHP)

Traditional centralised power stations lose much of the heat that is produced from the generation process. Apart from heat lost in the cooling towers, electricity is transmitted hundreds of miles by pylon, transformed and fed into local distribution networks, losing energy at every stage.

Woking has three small CHP systems using natural gas to generate electricity. The electricity is fed into a local network, minimising losses through transmission. The waste heat is used directly for heating. For example, the town centre CHP provides electricity, hot water and chilled water not only to the civic offices and a car park, but also to two hotels, a leisure centre, a night club and a conference centre. The whole system is self-contained and normally operates completely independently of the electricity grid.

A smaller CHP unit has been installed in one of the council's sheltered housing schemes. Again fuelled by gas, it provides heating and electricity to its residents and electricity to the nearby medical centre. This scheme also incorporates PV solar panels, which produce part of the electricity. Combining PV with CHP provides a better financial payback than could be achieved with solar PV alone.

The council's third CHP unit is based at the Pool in the Park leisure centre and also runs on natural gas. In this case, instead of an internal combustion engine driving a generator, hydrogen is stripped from the gas and fed to a fuel cell which produces electricity – and heat – directly. This is the only commercial fuel-cell CHP in the country and is the cleanest method of producing electricity and heat, with minimal emissions.

Small projects

Solar powered parking meters are found all over the country and Woking has been using them since 1997. They have no running costs and they are cheap to install because no trenches or cables are need to connect them to the electricity supply.

Woking now has solar-powered lampposts as well, which also have vertical wind turbines to boost the battery on dull days. Although they cost far more to buy than ordinary lampposts, the savings on installation and running expenses mean that they will pay for themselves in 2.4 years, which is very acceptable for a capital project.

Other projects included updating lighting systems to use Compact Fluorescent Lamps (CFL) instead of tungsten lamps and incorporating daylight sensors and presence detectors to turn lights off when no-one is about. Water is often forgotten, but is a high energy user and the introduction of waterless urinals has made important savings. Projects like these saved the Council approximately £50,000 per year just for the Civic Offices alone.

The Council is now preparing to replace all of its large scale computer monitors with flat screen monitors connecting to existing Thin Client server devices. These radically cut down power consumption as much of the computing power is provided by the network server rather than a by PC on every desk.. It has been calculated that this new approach will save approximately £400 per year per device due to a reduced need for power and a reduced demand for cooling. It will also save 250 grams of CO_2 per workstation.

Financial performance

Any public body has to manage public money with care, so while the councillors have been generally supportive of environmental measures they have only approved these projects on the basis that there is no increase to Council Tax. There is no requirement to save on council tax for environmental projects, but they must at least be cost neutral.

The Council's larger projects such as the CHP and PV installations generally have a payback of between 20-30 years. While the

Council's initial energy efficiency strategy was driven by the potential financial savings, its objectives in its sustainable and renewable projects were not purely financial. This strategy has now been embedded in the Council's climate change strategy to reduce CO_2 equivalent emissions; adapt to climate change; and promote sustainable development. Savings are calculated on a CO_2 reduction basis.

The climate change strategy also has a wider remit – covering the whole of the Borough's energy uses, services and environment. As at March 2005 the Council had saved 51% in energy and reduced CO_2 equivalent emissions associated with its own buildings by 79% since 1990.

As the Council moved towards larger projects it set up Thameswey Limited and Thameswey Energy Limited to manage the schemes. These companies also pursue energy and environmental projects anywhere in the UK. Projects are generally undertaken on a 20% share capital and 80% loan basis with returns of between 8% and 12%. Profits are re-invested within the Borough.

Net savings count towards the targets for local authorities set up by the government after the Gershon review into public spending.

Woking is not alone.

SOUTHAMPTON

Many operations concentrate on energy saving rather than specifically the use of renewables. Like Woking, Southampton has installed a CHP system, but it has an extensive district heating network and it was the first place in the country to make use of geothermal energy.[57]

There are two sources of geothermal energy. One, in places like Iceland, occurs where volcanic activity continuously pushes boiling mud and steam to the surface. The other is where heat has been

....

[57]Download case study from *http://www.managenergy.net/products/R124.htm*

gradually radiated out from the centre of the earth and trapped below the surface. Because this heat has been built up over the life of the earth it is not strictly renewable except over many centuries. Nevertheless, useful energy can be extracted for a number of decades.

After initial experiments in 1981, the Council set up a partnership with Utilicom, an energy management company, and drilled a well in the centre of the city. At a depth of nearly 2km, water was found at 76°. This is pumped up to the heat station and provides 18% of the energy required by the CHP system. The district heating system uses a network of 11km of pipes, and the insulation limits temperature losses to 0.5° per km. Heat is supplied to commercial customers as well as council properties and a separate pipe network delivers chilled water for air conditioning. The whole thing has been working successfully for some 20 years.

CHP is only part of the Council's approach to energy efficiency, cost saving and emissions reduction. The Council has a number of energy efficient housing improvement and building schemes in progress. It is working together with businesses and other organisations towards the adoption of a Green Transport Plan and already owns 28 alternative fuel vehicles and 13 cleaner emission refuse vehicles. Sixteen of the buses currently operating in Southampton are powered by natural gas and many of the old diesel buses are being replaced by cleaner versions.

At the other end of the country a different approach has similar objectives.

TEES VALLEY

In the North East of England the Tees Valley is promoting itself as a UK centre for renewable energy and waste management technologies.

Wind

The intention is not only to take advantage of the new technologies but to regenerate industry by establishing local manufacturing facilities for renewable energy equipment. For example, the region has permission to build one of the biggest wind farms in England. At present most equipment of this type comes from continental Europe, but the intention is to build the turbines locally. This will use local skills and develop into a business supplying wind turbines to schemes elsewhere in the UK and Europe.

Hydrogen

In the previous chapter we considered the use of hydrogen as a store for surplus energy generated by renewables. Tees Valley is in a unique position to put this into practice. Because the region's chemical engineering heritage has left vast underground caverns and a network of large diameter pipes, the plan is to use these as a ready-made system for the storage and distribution of hydrogen. There will be local investment in the development and application of the hydrogen fuel cells needed to return the energy in the hydrogen back to electricity.

Another application for hydrogen has been created by local firm Varitext. The company's traffic signs can be powered by a hydrogen fuel cell instead of the conventional diesel engine. This is presumably valuable in residential areas as the fuel cell is silent in operation. The signs are refuelled with hydrogen cylinders as required, but there is even one site which is close to autonomous. It has a solar panel which provides electricity to electrolyse hydrogen from rainwater (plenty of that in the Tees Valley) and the hydrogen can supply the fuel cell to provide electricity to the sign after dark. They claim that this is a more environmentally-friendly solution than the Woking solar/wind lamppost as there are no batteries to manufacture or eventually dispose of.

Biomass

In an area dominated by coal, a 30MW wood-fired power station will open in 2007. the unit will burn not only short-rotation willow coppice, but also offcuts and waste from demolition, construction sites and factories; sawmill waste and logs produced by thinning operations in local forests.

Biodiesel

The Biofuels Corporation has already built its first plant in the Tees Valley, to produce biodiesel from oilseed rape, palm and soya. A second plant is planned for the site, creating the world's largest biodiesel production complex. As we have seen in the previous chapter, biodiesel production on a major scale will require imports of suitable vegetable feedstocks. Palm oil in particular is likely to cause more environmental damage than the biodiesel ever saves.

So far these examples have involved organisations in the public sector or with significant government support, but the private sector is also taking steps towards efficiency and reduced emissions.

SUPERMARKETS

In January 2004, in partnership with Solar Century and partly aided by a government grant, Tesco installed a solar powered roof at a new petrol station in Hucknall, generating 20% of the energy used by the petrol station and preventing five tonnes of CO_2 being generated each year. Installation cost compared with the cost of the energy saved indicates that the project will pay for itself in about 30 years and the company is evaluating the potential for using solar panels to generate at least part of the power for in-store lighting.

The company has started installing CHP units and finds that these have a payback of six years. The potential for wind turbines and hydro-power is also under consideration.

Working with the Carbon Trust, J Sainsbury achieved a 20% reduction in carbon emissions in 2004/05, equivalent to 77,030 tonnes of carbon per year. The reduction in energy use equated to a cost saving of in excess of £8 million per year from the start of the initiative.

A number of Sainsbury supermarkets use lower fossil fuel energy, including wind turbines at the East Kilbride depot and at the Greenwich and Kingston stores. They too have installed solar panels on a petrol station and they have CHP systems in five stores.

Asda in Southampton is a client of the council's district heating system.

THE CARBON TRUST

The Carbon Trust has been established by the government to assist companies to reduce their CO_2 emissions. It does this to a large extent by advising companies on how to reduce energy consumption by using it more efficiently.

Armstrong World Industries, a manufacturer of floor coverings, was able to save 23% of its energy costs after two years. Much of this was achieved without investing in new systems or equipment. An energy audit revealed that power could be saved by turning things off when not needed, particularly at night and on Sundays and Bank holidays. It also identified areas which were heated unnecessarily. Staff training raised awareness of the energy issue among the workforce and ensured they understood the reasons and importance of keeping consumption under control.

Epsom and St. Helier University Hospitals NHS Trust was spending more than £1.5m on energy when it first contacted the Carbon Trust. The operation is spread over four sites and there were various systems for heating, lighting and air conditioning. Replacing old boilers and installing a computerised energy management system have yielded 10% savings. With more investment further savings may be achieved.

In manufacturing, the Carbon Trust worked with the BMW plant at Hams Hall, which was also spending £1.5m on power each year. Again, energy awareness among the workforce was important in reducing costs in what was already a very efficient plant. Together with improved control of the manufacturing services equipment this achieved savings of 15%, repaying the investment in less than a year.

WBB Minerals was spending nearly £9m annually, mainly on sand drying systems. As a major element of the company's costs, energy prices have a direct effect on the company's profitability. Improved controls and production methods produced savings of 8%, some £700,000.

TRY THIS AT HOME

So far we have considered alternative energy and energy saving at the commercial level. What can we do at home to improve energy efficiency and benefit from alternative fuels?

The most effective way of saving energy is using less of it. This is also the best way to save money. As somebody said, the greenest mile is the one not driven. You can look at alternative ways of generating electricity and we consider them below, but even if you get a grant you will generally NOT save money. Your energy will cost you much more than just buying it from the usual suppliers.

Insulation

The first and probably most obvious step is to make sure that your house, office or workshop is properly insulated. In fact, if this has not been done you will not qualify for any of the government grants for solar panels, wind turbines and so on. The benefits of insulating lofts and cavity walls, fitting double glazing and lagging the hot water cylinder are well known. If there are any properties out there without insulation – in fact a surprising number of cavity walls are still unfilled - there are still grants available. Ask your local council. You may even be able to get a grant to upgrade your existing insulation, for example by installing an extra layer in the loft.

Lighting

You can start saving energy by replacing some or all of your light bulbs with low energy units (Compact Fluorescent Lamps – CFLs). They are significantly more expensive than the traditional filament bulbs, but they last much longer and they use much less energy to give the same amount of light. The overall cost is much lower, so you will get more than your money back. For example, a 20W CFL gives the same output as a traditional 100W bulb. A CFL costs around £3 (less on-line) as opposed to £0.50 for a filament bulb, but a CFL lasts 8000 hours and a filament bulb only 1000.[58] The chart shows that the 80% saving on running costs repays the higher purchase cost several times over.

....

[58]The National Energy Foundation *http://www.nef.org.uk*

The reason that a traditional bulb is inefficient is that most of the power used goes into generating heat, not light. It has been pointed out that if you remove the heat from the light

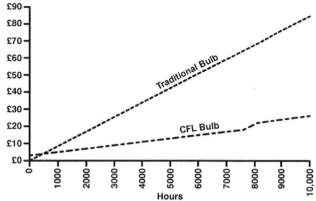

bulbs from a room then the central heating system will have to supply more heat to make up for the shortfall. It is argued that this will offset the saving from the low energy bulb to some extent. At least this will not be a problem in summer, and if you have an air conditioner it will not have to work as hard to overcome the heat from the lights.

There are some minor drawbacks with CFLs. Some people do not like the colour of the light produced, but aesthetics v economics has to be a personal decision. From a practical point of view, CFLs will not always work with dimmers or with other electronic devices such as time switches. Special wiring may be needed to make them work with security devices – passive infra-red motion detectors or daylight detectors.

Low voltage bulbs are NOT low energy, so those halogen spots you have in the kitchen are not saving you money!

Stand by for savings

How many items do you have on stand-by? Apart from the TV, video and satellite box you may have an alarm-clock radio, a microwave, your PC (together with printer and modem), portable telephones, a mobile phone on charge, and a clock on your oven. All of these use tiny amounts of electricity so it's difficult to believe the newspaper articles which frequently claim that if we did not leave things on standby we could close several power stations! Can a TV use nearly as much electricity when it's on standby as it does when it's in use?

I decided to find out and I carried out an experiment. This was not a scientific experiment and is purely anecdotal, but it does indicate that the stories are probably right. I obtained a consumption monitor.[59] This is a plug-in wattage and current meter and shows, among other things, how much power a particular device consumes. It shows that when I am watching my favourite programme my TV is using about 120W, or 0.12kW. When it is on standby, the meter tells me that it is using 16W, or 0.02kW. On the face of it the story is wrong – the set uses far more when I'm watching it than when I'm not. However, let's look at a 24-hour cycle.

One unit of electricity is 1kW/hour – using 1kW for 1 hour, 0.1kW for 10 hours or 0.01kW for 100 hours. I therefore consume one unit of electricity, costing roughly 15p, by running my TV for 8 hours 20 minutes. Actually, I rarely watch any more than 4 hours per day, which consumes 0.48kW/hr or just under half a unit. For the other 20 hours on standby the set uses 0.016kW x 20hrs = 0.32kW/hr. But this is two thirds or 67% of what it uses when I'm watching! So I spend 7p per day to watch TV and at least 5p a day not watching it. That's £18.25 per year to not watch TV.

Of course electricity is relatively cheap, which is why £18.25 per year is irrelevant for most people. Even if you add the cost of all the other items on standby, the digibox, the video, the phones, the microwave, the alarm clock and so on, it's not going to be much more than £50 per year. And £1 a week is not really going to upset anyone.

On the other hand, suppose that every household in the land has appliances on standby consuming a constant 50W. And let's suppose there are about 25 million households in Britain. That's a load of 1,250,000,000w or 1,250,000kW or 1,250MW. This is the same as the output of Torness nuclear power station – 1250MW. If we turned off everything on standby, we could close this station down. By comparison, the projected output of Europe's largest wind farm is 750MW, just over half what we use on standby.

So should we be building wind farms, nuclear power stations and installing solar panels, or should we just turn the TV off at the wall?

．．．．．．．．．．．．．．．．．．．．．．

[59] **http://www.brennenstuhl.de** *This unit is available widely on the web and also from*

Getting around

There are many books which preach the green lifestyle and urge you to walk, cycle or catch the bus. All these things will help you reduce your energy consumption, but generally they will take far more time and for some people they are just not possible. You can choose a smaller car. (Though you will not buy a hybrid if you believe research from the US that suggests that a 15mpg Hummer is much cheaper in terms of whole-life cost. This may be true at US petrol prices, but certainly not at 90p/litre. The study does not take relative carbon emissions into account.[60])

Much of our travel is dictated by our lifestyle, and many lifestyles are dictated by the fact that travel is possible. It's nice to live in the country or the suburbs, but does it make sense to spend millions of man-hours and millions of litres of petrol to commute to work each day? Does it make sense to spend all this time and resources to travel to a place where we can use computers and equipment like the ones we have at home? Of course you can't go to the dentist on the internet (yet); you can't print a newspaper or build a bridge without leaving home. You can only build social and business relationships face to face, but do we need to meet five days a week? Many business operations can be carried out almost anywhere, which is why we find call centres almost anywhere.

If your daily commute cost you £100, (that's £2000/month) how would you adapt? If someone blockaded the Middle East and petrol went up to £5 a litre –when you could get it – what would you do? Give it some thought. As the Boy Scouts say, 'Be prepared!'

Costly alternatives

Solar energy is free, the wind is free and hydropower costs nothing. On the face of it these are all excellent ways of saving fossil fuels, saving carbon emissions and saving money. The Energy Saving Trust[61] publishes cost-saving figures, but these appear to play down the actual cost of buying the equipment. Unfortunately, with the possible exception of heat-pumps, alternative energy systems will cost you far more than buying your electricity from your local power company.

....

[60]*Dust to Dust Energy Report – CNM Marketing Research Inc.*
[61]*The Energy Saving Trust* **http://www.est.org.uk**

If you go for micro-generation – solar panels or a wind turbine on your roof – it appears that you can make savings in two ways: by using the electricity you have generated or by selling any surplus back to the grid. Indeed, it is a condition of the grants that you sell your surplus back to the grid.

According to the Energy Saving Trust, you will sell your energy for less than you can buy it back. This is not unreasonable when you consider that you are just generating electricity. The power company is generating it, transmitting it along the pylon routes, transforming it and distributing it to consumers. They also have to charge you VAT, and, of course, consumers can't charge VAT back to them.

For a number of reasons, domestic alternative energy installations are less cost-effective than commercial ones. A major reason is that the greatest economic benefit comes from using the energy you produce yourself, rather than selling it to the power company and having to buy it back later at a higher price. Many people are away from home during the day and so are not using the energy from their solar panels. By contrast, commercial installations can be more efficient because commercial premises have highest demand and use the most electricity during the day, at the same time as photovoltaic (PV) solar panels are likely to be at their most productive.

Of course, you could have batteries at home to store your electricity, but this means extra cost and extra control electronics and you could never generate and store enough to provide for all your requirements.

With solar hot water there is no electricity involved, so nothing to sell back. Thermal stores can increase the heat retained, but also increase the cost. In any case, you only get the full benefit if you use all the hot water you generate.

Wind turbines have much the same cost profile as PV panels, although they can produce electricity at any time of day or night as long as there is a wind. There needs to be enough wind!

Let's look at some specific costings. There are many different tariffs, involving standing charges and no standing charges, night rates, direct discounts and loyalty bonuses, so it is very difficult to say exactly what electricity costs the UK consumer. As a reasonable approximation of current prices (December 2006) I have taken 15p for the normal rate and 4p for off-peak.

SOLAR ELECTRICITY (PV PANEL)

A solar panel is a bit like a flat TV but absorbs light instead of emitting it and turns 14-17% of the sunshine that falls on it into electricity.

The property remains connected to the grid in the normal way, so when there is insufficient solar electricity, the shortfall is taken from the normal supply.[62] When you have a surplus you can sell it back to the grid, but you will receive a lower price than the normal domestic price paid to purchase electricity. As an example, Ecotricity are currently offering 4.5p per kWh, compared with the 15p that a consumer would normally pay.

If we assume that:

◊   An installed system producing 2kWp (kilowatts at peak output) will cost £10,000 - £20,000

◊   In 2005 there were 1545 hours of bright sunshine in England (Scotland 1188, Wales 1435)

...then the economics of the system look like this.

Cost of installing solar panels (2kWp)	£15,000
Government grant[63]	(£6,000)
Annual hours of sunshine	1500
therefore Electricity generated per annum	3000 units
Saving 3000 x 15p	£45

....

[62]*Some systems use batteries, but since batteries generally store electricity at a much lower voltage than the mains there are issues of transforming and rectification which imply conversion losses and a need for additional equipment.*

[63]*Low Carbon Buildings Programme:* **http://www.lowcarbonbuildings.org.uk/home/**

On these figures it would take 20 years to recover the £9,000 investment (£15,000 less £6,000 grant) – no savings would be made until year 21.

This will only be the case if the consumer uses all the power generated and sells none back to the grid – an unrealistic situation. Suppose 25% is sold back. The saving then looks like this:

Saving 2250 x 15p	£338
750 x 4.5p	£34
Total saving	£372

The payback period is now 24 years: no savings for the consumer before year 25, and by this time the panels would be close to worn out.

This scenario also ignores the cost of borrowing the money to purchase the solar panels. Even if you are prepared to ignore that cost because you have some money readily available, you are foregoing the interest you could otherwise have earned on your £9,000 After 20 years invested at 5% your £9,000 will be worth £23,880. If you invest it in a solar panel you will not even have broken even.

We all expect energy prices to rise, so the table below shows the break-even period assuming annual price increases of 7% (the savings will be worth 7% more) and also assuming that it costs 6% to borrow the money. Payback takes 23 years.

If energy costs rise by 7% per year, what costs you £372 in 2007 will cost you £456 in 2010 and £639 in 2015. Recent increases in 2006 have been of the order of 20%. If that continued on an annual basis you would pay off your solar panel within 12 years. However, energy costing £372 in 2007 would be costing you £1,600 in 2015.

If you have to pay 6% interest on your investment but there is no increase in energy prices, you will never ever save enough to pay for your panels.

In summary, domestic PV panels are not financially viable. You'd be better off investing your money and using the interest to buy some low-energy light bulbs.

GENERATING POWER AT HOME - COSTS AND BENEFITS

	Installed Cost	Grant	Net Cost	Annual Output Kwh	Value of Annual Output	Payback			Carbon Savings - tonnes per year
						Simple payback (years) - no borrowing, no fuel price inflation	Payback (years) if money borrowed at 6% and fuel price inflation = 7%	Payback (years) if money borrowed at 6% and NO fuel price inflation	
Solar Panel (PV) 2kwp - electricity	15,000	6,000	9,000	3,000	372	24	23	never	1.53
Solar Panel - hot water	4,000	400	3,600	2,000	60	60	never	never	1.02
Wind Turbine	2,600	500	2,100	660	119	18	17	never	0.34
Heat Pump (8kw ground source)	8,000	1,200	6,800	10,240	307	22	21	never	5.22

SOLAR HOT WATER

Domestic water heating schemes consist of solar collectors, a preheat tank, pump, control unit, connecting pipes, the normal hot water tank, and backup heat source such as gas or electric immersion heater. The collectors are mounted on the roof and heat the water tank via a fluid circulated between the collectors and the tank. The overall area of the panels is typically 3-4 square metres.

The panel is usually made of metal and installed in a glazed box to insulate it against the outside air. Costs vary and even a DIY kit costs more than £2,000. A professional installation, possibly using copper pipes for improved conductivity or vacuum-sealed glass tubing, could be as much as £5,000.

Solar water-heating is an eco-friendly solution – no CO_2 emissions, no fuel – but, as shown in the table, the energy saved will never cover installation costs; (and if you move house you have to pay the grant back!)

Solar hot water panels cost less than the PV panels which generate electricity, because they simply heat water. Even on dull days they have a heating effect and can reduce the gas or electricity needed to bring your hot water to normal temperatures. They can be used in conjunction with a thermal store, but this is only worth doing if you are going to use the extra hot water that this will produce. If you are seriously environmentally-conscious you may be cutting back on your water use anyway. An electric pump is usually needed to circulate the water from the solar collector to the storage tank, and the energy needed to run this should be offset against the savings. DIY designs for solar hot water systems are widely available and can be much cheaper than professional products, but there are no government grants for DIY.

Once again, at the domestic level solar panels will not pay for themselves. This was underlined by an advisor at a recent presentation by the Energy Saving Council. "If you come into some money," he said, "you really ought to consider a vacuum-sealed system" – thus making it quite clear that a solar panel will only pay for itself if it's free!

A WINDMILL ON EVERY ROOF

Sounds great, doesn't it? Everyone has a roof, and the wind is free! Everyone's doing it, even David Cameron, leader of the Conservative Party. You can get windmills from more and more places, including your local B&Q DIY store. And in almost every case it will be a total waste of money.

The UK has 40% of the wind in Europe, but a wind turbine needs a strong, steady flow. This is available on hilltops, coasts and out at sea. It's not available in urban areas where buildings and trees break up the airflow and cause turbulence and eddies. This makes the turbine swing backwards and forwards and speed up and slow down. Turbine output is directly proportional to speed. Most appliances need a steady supply of current – electronics are particularly sensitive – so hopefully the control systems will smooth all this out by importing and exporting to the grid from second to second. But all this erratic motion is not very good for the turbine – it will wear out much more quickly if it's not in a steady breeze. And then there's the noise it makes. Maybe you will put up with it for the sake of the environment (though it's unlikely that your turbine will actually create any net environmental benefit) but how will your neighbours react to the noise?

A popular wind turbine will produce 400w at a wind speed of 16 metres per second. Another model will produce 1kW at 12.5m/sec, though only if the wind is steady, not gusting, and it cuts out at 14m/sec. Neither unit will provide more than part of your electricity requirement.

The average wind speed across the UK is only about 4m/sec.! You can find the average wind speed for your location from the DTI database.[64] You will see from the tables that it gets faster the higher you are above ground level, but even a 45m tower (assuming you get planning permission) is unlikely to get enough speed in urban areas. Roofs, chimneys and trees all interrupt the flow and cause eddies and gusts. Here is a quotation from an installation guide from one of the manufacturers:

....

[64]*The British Wind Energy Association gives a guide to accessing this data*
http://www.bwea.com/noabl/

"Some parts are heavy, and are mounted high in the air where they pose the potential of becoming a 'falling hazard' in fault conditions or in high winds, and as such, every effort should be made to keep a safe area free from people, animals, buildings and vehicles around your turbine at all times.

"You will be required to maintain your turbine … … this will mean tilting your tower down to ground level on occasions, which should be carried out with the assistance of at least two other fit and able people.

"Your turbine should NEVER be run without a load connected to it. This could result in the blades breaking, your tower collapsing, bearings being destroyed prematurely, electrocution, personal injury and even death."

And as you can see from the table, your turbine is unlikely to ever pay for itself.

GROUND SOURCE HEAT PUMP

A ground source heat pump has an advantage over solar and wind systems in that it works on demand and does not depend on the weather. A ground source heat pump works in exactly the same way as the heat pump is used in an ordinary fridge. Using pipes buried in the garden it extracts heat from a large area below the surface.

Only a few metres below the surface the temperature remains pretty constant throughout the year. In the same way as a fridge takes the heat from inside the cold compartment and the radiator on the back gets hot, so the heat pump takes heat from the garden and feeds radiators and possibly a thermal store in the house. The pump takes a lot of relatively low-temperature heat and converts it to a smaller quantity of high-temperature heat. The low-temperature heat comes from a large area, so there is never enough taken to freeze the ground.

As with the fridge, the heat pump needs electricity to operate, but for every 1kWh of energy used by the pump the system harvests 3kWh or 4kWh of heat. It can be switched on whenever required, and if it is installed with a thermal store it can run at night and make use of off-peak electricity, which is currently around 4p per unit rather than 10p-18p on peak tariff.

On the other hand, since the heat pump produces heat, not electricity, the savings and payback period must be calculated in relation to alternative costs of heat. You have to balance the costs of off-peak electricity, which is only about 4p per unit and gas is about the same, with the cost of installation and the amount of usage a ground source heat pump will get during the colder months of the year.

A ground source heat pump is likely to be a good solution where there is no gas supply, particularly if it is used to replace an oil-fired system.

MICRO HYDRO-ELECTRICITY

Each micro-hydro installation will be different, so it is not possible to make generalisations about the cost-effectiveness of the system. Few people will be able to use micro-hydro, but if you are fortunate enough to have a fast-flowing stream on your property you may be able to harness it to generate electricity. You should ask your estate manager or butler to make enquiries.

Seriously, micro-hydro will rarely be an option. You may be interested to know that the very first domestic hydro-electric scheme in the world was constructed at Cragside in the 19th century. This was the country house of Lord Armstrong, who built ships, bridges and armaments in nearby Newcastle. The estate is now a National Trust property and open to the public. It is called Cragside because it is on the side of a crag. This provides the ideal location for falling water to drive the turbines.

BIOMASS

Biomass is the posh name for wood and other organic products and biomass is popular because at first sight it is carbon-neutral. This means that it gives off CO_2 when it is burnt, but growing another plant to replace it will soak up the same amount of CO_2, so there is no net increase in global CO_2 levels. Of course, this only works if a tree is planted for every tree that is burnt, and it takes no account of the carbon emissions from the machinery needed to plant, maintain, harvest, transport, process and deliver the fuel. Biomass and other biofuels may be a green choice, but they are never carbon-neutral. Some have even calculated that in certain circumstances biofuels are carbon-negative, since the whole process emits more carbon than the biofuel saves.

Tars and other pollutants are usually emitted when wood is burnt in a domestic stove. To counteract this there are now some technically sophisticated stoves which burn at extremely high temperatures, consuming most by-products and creating very little ash. They can even have electronic remote controls, automatic lighting and room thermostats.

The disadvantage of such units is their cost and the fact that they generally use wood pellets which involve an additional manufacturing process and long-distance transport. Every installation is different and there is a wide range of fuel sources, so the cost-effectiveness and the greenness of each biomass project must be assessed individually.

GOVERNMENT GRANTS[65]

The scenarios we have looked at all involve government grants. Even with these grants, domestic generation does not make financial sense. You will not save any money in a realistic time. Even the figures in the table are over-optimistic as they do not allow for the cost of maintenance or repairs or VAT on the original purchase. Solar panels may last 25 years, but heat pumps and wind turbines almost certainly will not. This means that you will never get your money back before you have to replace them, let alone make savings.

....

[65]The Low Carbon Building Programme operated by the DTI provides details of the grants available for the various different systems.
http://www.lowcarbonbuildings.org.uk

The grants themselves make good headlines, but let's look at exactly what they amount to. The Low Carbon Buildings Programme has £10.5 million available to support domestic and small community projects. This sum is to support PV panels, wind turbines, solar hot water, heat pumps and micro-hydro. If, for example, we allocated the whole amount to PV solar panel projects there would be enough for 1,750 households out of the 25,000,000 households in the UK. These projects would save 2,678 tonnes of carbon emissions each year. Compared with the UK's total annual carbon emissions of 150,000,000 tonnes, that is 0.002%.

In 2006, the low Carbon Buildings Programme grants ran out well before the end of the year. The £10.5m annual grant fund is roughly equivalent to what we spend on petrol and diesel in the UK in half a day, every day.

The UK's Kyoto target is to reduce carbon emissions by some 20m tonnes by 2012. Although people who live in flats cannot do anything, if everybody else installed a windmill and a solar panel on the roof we would go a long way towards achieving this target. It would cost about £400,000,000,000, or £20,000 for every house with its own roof. Carbon would be reduced, but consumers would see no savings for over 20 years and it is blindingly obvious that very, very few households would bother (or could afford) to make the investment.

In view of this it is difficult to see why the government is investing in not only the grants, but also in the cost of running the Low Carbon Buildings Programme and the Energy Saving Trust. There will be no measurable benefit to anyone. By comparison, closing coal-fired power stations and replacing them with gas units would make a substantial difference, and replacing them with nuclear stations, which have no CO_2 emissions in operation, would cut CO_2 emissions substantially more.

This, of course, raises questions about importing gas and the public's worry about nuclear safety.

ADDING IT ALL UP

There are no benefits from micro-generation by individual households. There are problems with renewables on a commercial scale – mainly that they will never be big enough to satisfy a useful part of the demand. Biofuel production on any significant scale means imports, and the raw material for biofuels can distort food markets and create more environmental damage and carbon emissions than they avoid.

On the other hand the Carbon Trust and pioneering local councils have shown us that significant energy savings are possible. One of the most promising approaches is combined heat and power (CHP) linked with solar energy and possibly ground-source heat-pumps, at a community level. Domestic micro-generation is generally too small to be economic, particularly as the domestic demand for power does not coincide with available supply. However, if councils or other organisations can set up CHP and renewable systems to supply mixed estates, with both commercial and domestic properties, the energy can be consumed as supplied, minimising the transfer to and from the national grid. We're not talking only of electricity here, but of heat which would be lost if power came from a remote station and transmission losses which are minimised when power is transmitted within a small local area.

We can improve the efficiency of energy use and reduce the amount we need, but we cannot eliminate it altogether. Have we any viable alternatives to the consumption of fossil fuels which sustains our present lives? If not, how can we reduce consumption of what we have left?

How long does it take to install insulation? How long does it take to install district heating, to revise building regulations, to introduce a wide-ranging grant scheme, to upgrade public transport and stop building roads?

How long before we suffer serious economic collapse because we can no longer afford to run the infrastructure we all rely on?

Action This Day

WILL CLIMATE CHANGE YOUR LIFE?

What do we do now?

CHANGING THE CLIMATE

The problem with climate change is that there is so much that we don't know. The problem with energy supplies is that nobody seems to want to know. There is much high-profile comment on the consequences of climate change, but very little on Peak Oil or the risks of an energy shortage.

We'll look at climate change first.

In November 2006 Johnny Borrell was one of several thousand people who joined the Stop Climate Chaos march on Trafalgar Square.

"Together we send a message," he said. "Together we can Stop Climate Chaos. Together, we count."

The whole event and comments like these emphasise the frustrations, the enthusiasm and the incomplete knowledge which surround the whole issue. There is no shortage of campaigners and pressure groups that highlight the dangers, the threats and the things that are going wrong with the climate. Some are very weak on solutions – some are dogmatic and single-minded on what must be done. The Climate Change Camp in September 2006 marched on Drax, Britain's largest coal-fired power-station in a symbolic attempt to close it down.

Al Gore has written his book *An Inconvenient Truth* and released a film of the same name which both warn of dire consequences but have little to say about what we do. Indeed, the film says nothing about solutions and actions, waiting until the final credits before slotting in some suggestions like 'Recycle plastic bags', 'Turn your thermostat down' or 'Drive a smaller car.'

A much more valuable though less prominent book is George Monbiot's *Heat*, where the author goes into meticulous detail about the science of climate change, quantifies the actions needed to correct the situation and examines how each possible solution could contribute to the overall need.

Sadly the popular public perception of the dangers and solutions to climate change is still simplistic and ill-informed. It seems to take the view that:

1. Global warming causes dangerous climate change

2. Carbon dioxide causes global warming

3. Human activity produces excessive carbon dioxide

Therefore

4. We must reduce our carbon dioxide emissions

5. Dangerous climate change will stop

As stated at the beginning of this book, ***"Cutting CO_2 now will NOT stop global warming or its immediate consequences."*** What we do today, however, will have a crucial effect on our children and grandchildren for the rest of the century and centuries to come. For most of us, it will affect the climate for the last years of our lives.

UNITED NATIONS

There have been many calls to action, but many have gone unnoticed and unheeded. At the UN's Rio Earth Summit as far back as 1992 a Framework Convention on Climate Change was agreed, with the following objectives:

◊ To return greenhouse gas emissions from developed countries to the 1990 levels by the year 2000

◊ To aim for long-term stabilization of the climate while allowing ecosystems to adapt, protecting food production, and a sharing sustainable economic development.

In fact, by 2000 emissions were up by 10%, and this could have been very much more except for the collapse of the Soviet Union which meant that emissions from that region declined by 40%.

Certain greenhouse gases are under control and CFCs are an excellent example. CFCs are chlorofluorocarbons, highly efficient gases used in refrigerators to absorb and remove the heat. CFCs were also used to fill the bubbles in insulating foam in refrigerators. In the 1970s it was discovered that CFCs released into the atmosphere were breaking down into chlorine and other substances and damaging the ozone layer. The hole in the ozone layer let in harmful ultra-violet (UV) rays which could cause skin cancer among sunbathers and people working out of doors.

Under the Montreal Protocol of 1987 the nations of the world recognised the danger of CFCs and agreed to discontinue their production and use. Although their effect will continue to be felt for a hundred years, the protocol has been largely successful. HCFCS which replaced them are also to be eliminated, and they also will have the residual effect for 100 years.

Ten years later the 1997 Kyoto Protocol set specific and legally binding greenhouse gas emission targets for each country, to be achieved by 2012. It also introduced new ways in which this could be achieved, including joint implementation between developed and less developed states, the clean development mechanism and emissions trading. It was estimated that these measures could cost a developed nation up to 2% of GDP. Nearly all nations ratified the protocol. As we have learnt in Chapter three, the United States and Australia did not.

STERN WARNING

In late 2006 Sir Nicholas Stern, formerly of the World Bank, produced his review: *The Economics of Climate Change* for the British government. Sir David King, chief government scientist, had already identified climate change as a greater threat to our survival than even terrorism. Launched by the prime minister and the chancellor, the Stern Review re-emphasised this view and called for immediate action. In November 2006 a climate change bill was incorporated in the government's programme for legislation but was not expected to have detailed targets.

There is much in the Stern Review that is not new. This is not a criticism. Stern is an economist, not a climatologist, so he has built his economic review on the climate science established by others.

The 700-page report concludes that we need to act now to deal with climate change, since action in the future will be far more expensive than immediate action. The figures picked up by headline writers were an immediate investment of 1% of Gross Domestic Product (GDP) for the whole world, to avoid a future cost of 20% of GDP per year in the future.

In Stern's view we (that is all the nations of the world) need to achieve this by controlling greenhouse gases. Stern believes we can stabilise greenhouse gases at 550ppm[66] (and that the associated temperature increase will be sustainable) by cutting man-made greenhouse gas emissions by 25% by 2050 and by 80% by 2100. The cost of this will be 1% of GDP or some $300,000,000,000.

This means that although economic growth and greenhouse gas emissions have both steadily increased since the start of the Industrial Revolution, emissions in 2050 must be 25% lower across the whole world than they were in 2000, even though the global economy is expected to be 3-4 times as big by then. If the global economy in

....

[66] *The Stern Review talks about greenhouse gases: CO2, methane and the rest. Elsewhere in this book I refer to CO2 concentrations alone, so the values are lower. However, a letter in New Scientist (6th January 2007) suggests that Stern has confused these definitions in setting some of his targets.*

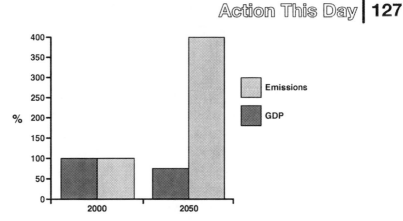

Can we achieve sustained growth and still cut emissions?

2050 is four times the size of what it was in 2000, the emissions per unit of GDP must be less than 20% of what they were in 2000. (See chart)

How does Stern think reductions will be achieved? The review identifies three areas for action: first, reduced demand for emissions-intensive products and services.

Reducing Demand

Taxation will reduce demand, and taxation was seized upon by the newspapers as the Stern Review was published. They predicted thousands of extra pounds for each family. If taxes are imposed; if governments feel able to do this, thousands of pounds will be the likely cost of the status quo. If people pay the taxes the policy will have failed. If people pay the taxes because they have no alternatives – no public transport, no other ways of heating their homes, no support for insulating their properties – they will pay them cynically, disbelieving that there is a climate crisis, seeing this as just another way of raising taxes.

Absorbing CO_2

Stern's second recommendation is to safeguard the natural absorption of CO_2. The Kyoto agreement has already permitted carbon emissions to be offset by 'carbon sinks'. A carbon sink is an area that absorbs carbon dioxide. Growing forests fulfil this role as plant life needs CO_2

to grow. While increased afforestation is not the whole answer to carbon dioxide reduction it can make a significant contribution.

On the other hand, if land which was previously desert or snow fields is planted with trees it can affect the earth's albedo, which is the earth's contrast or 'black-and-white-ness'. In other words, the light colour of deserts or snowfields reflects the sun's heat and light, whereas the darker green of trees will absorb the heat. So the carbon sink effect of a growing forest is offset to some extent by the warming effect of the darker, heat-absorbing colour.

It has been suggested that while nations get credit for new forests, they are not penalised for cutting forests down so they could get a credit for new forests while the net size of their total forest was getting smaller. As well as altering the albedo, deforestation emits CO_2 as the trees are burnt or rot. Brazil's immediate and firm response to the Stern Review was that other nations should leave the Brazilian rainforest in the safe hands of the Brazilian government.

Low-carbon solutions

To achieve stabilization requires anthropogenic (man-made) emissions to fall to a small proportion of the present level. That means that everything we do that emits CO_2 must fall to a small proportion of the present level – a small proportion of car use, air travel, electricity consumption and so on.

Stern recommends developing low-carbon solutions for power generation, heating and transport. These are technologies which we could develop in the developed nations and provide to the poorer nations, allowing them to develop in clean, low-carbon, resource-efficient ways.

THE TECHNOLOGICAL QUICK FIX

The precautionary principle requires us to take action now in case things go wrong, and Stern and many others are urging us to cut CO_2 emissions immediately. The intention is to avoid surprises.

The alternative is to simply hope for technical fixes and a number of suggestions have already been made. For example:

We could put mirrors into orbit to reflect back some of the sun's heat

We could put dust in the atmosphere, which would have much the same effect

We could seed the clouds to modify rainfall patterns

We could sow the oceans with iron compounds to stimulate the growth of plankton which in turn would absorb carbon dioxide.

We could do all these things, but nobody can be certain whether they will have exactly the desired effect or whether they will have undesired side effects. And there's the whole issue of political control. Who controls exactly how these measures work and who benefits the most from them? The dilemma is that in the early stages at least, it is the poorer countries that will suffer the most from the effects of climate change while the richest countries, with the greatest resources, include the greatest sceptics who see no need to do anything.

GROWTH

Amongst those who are sceptical of the challenge of climate change, much scepticism is based on the belief that the costs of proposed action will destroy economic growth. Stern, on the other hand, says "...tackling climate change is the pro-growth strategy" - because if we do not tackle it the consequences themselves will destroy economic growth.

How will growth be sustained in the face of these cuts? Stern predicts that:

◊   There will be expanding markets for low-carbon energy products

◊ Financial markets will become involved in carbon trading, financing clean energy and increased insurance business.

◊ As we move towards a low-carbon economy, inefficiencies will be reduced and governments will discontinue the subsidies which they have been paying to inefficient and unsustainable energy producers.

His appears to be, essentially, a macro-economic view: that given the right economic environment the markets will develop the right solutions. Stern is not a scientist, technologist or engineer: he seems to be simply hoping for the best. To be fair, he does not simply expect unregulated markets to provide the answer. Carbon pricing is a key part of his recommendations.

Carbon pricing is a system where factories, businesses, even individuals must pay for the carbon they emit as a result of their daily activities. Everyone has a basic allowance; if they use more they pay a surcharge, but if they use less they can sell their surplus to heavier carbon-users. At an international level, industrialised countries can buy carbon credits from less developed nations with credits to spare. The idea is to price carbon-emitting technologies out of the market over time, and to provide an income for developing nations as they sell their credits.

Technology policy is the second strand of the move towards a low-carbon economy. Governments are to fund research and development of new means of power generation, heating and transport. Carbon capture is an important area. It will not be possible to stop using fossil fuels for energy in the short or medium term, and fossil fuels produce CO_2. Carbon capture is a technique for removing the CO_2 from the products of combustion and burying it deep in abandoned coal mines, spent oil wells or at the bottom of the sea. We can also expect research into renewables such as tidal turbines, and the search for nuclear fusion continues. Our most important need is for an oil-substitute – a compact, portable store of energy for transport – but nothing seems to be in view.

'Removal of barriers to behavioural change' is another recommend-ation. It means training people and changing attitudes so that they accept that the risks and dangers are real, that the solutions are realistic and the benefits are worth the sacrifices. This must be the most difficult thing for any government or organisation to achieve.

The closure of much of the UK coal-mining industry has already shown how changes affecting a single group in society can cause anger, dissent and violence. If there are fewer cars, less jobs in servicing and repairs, in road construction and in fuel delivery, there will be unemployment and discontent. If the changes bring no apparent benefits people are going to revolt. Currently, the only true promise is that things *may* be less bad in 2050 if we act now. Michael O'Leary, head of low-cost airline Ryanair, will certainly make a battle of it. Commenting on the Stern Review he said: "A lot of lies and misinformation has been put about by eco nuts on the back of a report by an idiot economist."

DISPUTE AND DISINFORMATION

Firm conclusions cannot be based on unstable foundations, and Sir Nicholas Stern does not deny that both the costs and the impacts of climate change are 'subject to important uncertainties'. It is hardly surprising that this should be the case, since we are talking about global weather for centuries to come.

The problem is that there are groups with vested interests whose objective is to preserve the status quo and will seize on any uncertainty to question and deny the truth of climate science. It is suggested that these lobbyists are already attacking prominent scientists in order to discredit them and cast doubt on their contributions to the IPCC Fourth Assessment Report, due in 2007. This leads to two problems. On the one hand, the pro climate change lobby may defensively over-state its case, but in doing so frighten off the members of the public that it is trying to convince. The other danger is that pro-climate activists become dogmatic and reject any possible criticism, stifling true scientific debate. This is particularly dangerous where the problem is so immense and the risks are so high.

Informed criticism

Bjorn Lomborg, the Skeptical Environmentalist that you might remember from Chapter three, has criticised a number of aspects of the Stern Review and once again risks being shouted down by the eco-activists. The truth is that if we are going to commit massive amounts of money to addressing climate change we need to know exactly what we are doing and we need to keep reviewing the situation.

You will remember that Lomborg suggested in his book that we would be better off restricting our expenditure now and waiting until we had better technology and more resources in the future. Interestingly, he said we should consider spending no more than 4% of GDP – which is four times what Stern recommends.

Lomborg is prepared to accept that climate change is real and that it is partly caused by human greenhouse gas emissions, but he criticises Stern for concentrating solely on greenhouse gas reduction as the solution to the problem. He believes that sensible precautions against the consequences of climate change – flood defences, strengthened buildings in hurricane areas and so on – would reduce the serious future consequences that Stern uses to justify immediate action on CO_2.

He accuses Stern of cherry-picking the data so that he can make a clear choice between cheap action now and unsustainably high costs in the future. He believes that Stern has overstated the costs of damage in order to create a headline figure in support of his position. Equally he has understated the costs of action now, relying on market forces to drive technological change to deliver solutions.

Stern himself warns of risks and uncertainty – he could do little else. His detailed report shows a range of possibilities but his executive summary is necessarily brief and has yielded headlines out of context. Firm conclusions need firm foundations.

THE PEAK OIL PROBLEM

Let's not deny climate change. Let's not deny that the potential consequences could be far worse than anything we have ever experienced. But let's not overlook the problem of Peak Oil, which could change our lives as seriously and much sooner.

Throughout our lives we in the Western world have enjoyed limitless energy. Each year our demands have grown and each year supplies have met our needs. We take it for granted that the energy we want is there for us, because it always has been.

In a generation we have changed our lifestyles so that we leave lights on all day, at least in offices or shops, regardless of rain or shine outside. We commute far farther than our parents did: as new motorways are built the traffic expands to fill them. Almost anyone can afford a car, the petrol is always available and the real cost of motoring has fallen in recent years. We heat our homes to higher temperatures than we ever did, because we can and because it's cheap. As these things go on, we believe that we have a right to do whatever we can do. Where there are artificial constraints, like taxes, there are protests, and the 2000 fuel protest shows that governments can be made to back down.

Suppose Peak Oil is real. Suppose oil really is starting to run out. In fact few deny that it's real; some just prefer to believe that it's away in the far-distant future. There are three possible scenarios: a gradual change where prices go slowly and steadily up, a catastrophic reduction in supplies with doubling or trebling prices and shortages, or no real change – with supply matching demand for the foreseeable future.

Little by little

If prices go up gradually people will grumble and complain. Studies have shown that consumers will accept a substantial rise in petrol prices – even 50% or 100% before they would ever give up their cars. They will complain about increased heating costs and will demand action from the government. The government has already reduced petrol tax and has cut VAT on electricity to help people heat their homes. The government may pay lip-service to cutting carbon emissions, but no government has had the courage to tax people into changing their energy use or to invest in alternatives.[67] If oil prices continue to steadily rise, governments could offset the effect by cutting fuel taxes. They could justify this as a measure to avoid the economic depression that rising fuel prices would cause.

If prices go up gradually, governments will eventually have no fuel taxes left to cut, and they will already be unpopular because of the other taxes they will have imposed to make up the shortfall. At this point fuel costs will begin to bite.

More and more people will find they cannot afford to commute. They will find that houses remote from places of work will fall in value. Children will once again find themselves going to schools within walking distance. Will we work at home? Some can, some can't. You can't get a haircut or visit the dentist on the internet. If it is too expensive to travel to the supermarket will the corner shop re-open? Or will we all shop on the internet and rely on one van to deliver supplies to dozens of homes? Will we hire a car and take our holidays in the UK once high-cost aviation fuel ends cheap flights? The fact is that if we can only afford far less oil, gas and electricity our lives will change dramatically.

Surprise!

If a sudden oil shortage appears without warning, it will come to unprepared nations as more than a surprise. OPEC caused the oil shock in the 1970s and the consequences were severe, but OPEC was able to resume production and release the pressure on the West. Since that time the Middle East retains the largest oil reserves in the world, the West has increased its reliance on them and the Middle East has become less stable. Politically, states in the Middle East could disrupt supplies again; though western military powers would do everything they could to prevent this. Russia is becoming an

....

[67] *Yes, I know there are government grants for wind turbines and solar panels – as much as £10.5m in 2006 and all used up well before the end of the year. By comparison, we spend about £40m on petrol and diesel in the UK, every day.*

increasingly important oil and gas supplier and may be influenced, though cannot be controlled.

If the political threat can be managed, geological and practical threats could be much more severe. While output is close to its peak and demand continues to grow, price levels and supplies come under increasing pressure. Short-term problems will have increasingly severe effects. For example, in 2005 Hurricane Katrina disabled a number of oil rigs, leading to supply problems and price rises. Corrosion of pipelines in Alaska led BP to reduce operations, again affecting supply and pushing up prices.

These have been short-term and manageable, but it is not impossible that a major oilfield should suffer a geological collapse and simply stop producing oil overnight. Maybe a terrorist could bomb a well and start a fire that would take weeks to extinguish. This is the nightmare scenario – where immediate global shortages occur and the west is unprepared. This is where all the things we could do to adapt to a slow rise in prices have to be done at once, which is clearly impossible.

A sudden catastrophic failure of supply will finally make people realise that oil is limited: realise that if we use it now there will be none for tomorrow: not just none for our grandchildren, none for our children and none for our own old age. Markets will realise the true value of oil – as a limited raw material for plastics, fertilisers and finished goods. The price will reflect the fact that we cannot just burn as much as we like to get ourselves from A to B, or use it for one-trip plastic bottles and bags.

The day the oil runs out, or at least the day when oil suddenly becomes scarce, is probably the day when martial law is imposed. It's the day when the authorities take over the distribution – and rationing - of food, when they control who gets the petrol and when they manage the distribution of electricity. That day will be the first day of a whole new era, when people find they have to change their whole lives to live within the energy resources we have, not the unlimited energy we have always enjoyed.

ACTION THIS DAY

What can we do? Simply, be prepared. Governments can achieve far more than individuals, but only if there is the political will. At the individual level the best plan is to be aware of what is going on and be ready to adapt. Don't invest in wind turbines or solar panels for your house unless you've got a lot of money, you're prepared to cut your energy usage by 80% or 90% and you spend most of your time at home. Do choose a house in a location which will minimise your journeys – to work, to school, to the shops. Insulate your home and be prepared for travel, electricity and gas to cost far more and go on costing more.

You can avoid food miles by buying local produce, you can recycle and re-use, you can think hard about buying anything new. Some families already choose to do these things, though the effect on the environment and resource use is probably pretty negligible. If an oil shock changes our lives we will all have to do all these things: there will be few choices left.

Hope for the best - and expect the worst – is a prudent approach to life. (Governments are certainly good at the first bit.)

Driving On 8

WILL CLIMATE CHANGE YOUR LIFE?

Adapt and survive

If I am right, and energy shortages occur in the next few years, the effect on our lives will be dramatic. If I am wrong, but the government decides that we must cut carbon emissions at all costs, the effect on our lives will be dramatic as well. You can use this chapter as a guide to how the changes will affect you, or as a guide to how you can change your lifestyle so the changes come as less of a shock.

GREEN AT HOME

Using less energy will mean that your lifestyle is causing less CO_2 to be released to the atmosphere. Using less energy is a good way to get used to the fact that energy is likely to become scarce and expensive sooner than most people believe and sooner than most politicians admit.

Global warming is real. The problem is that we do not yet know how real or how soon or severe its consequences will be in terms of climate change. That is no justification for doing nothing, or relying on technology or future generations to solve the problem. Equally there is no justification for vilifying minorities, cutting CO_2 at all costs and ignoring all other approaches to the problem.

UNDERSTAND THE PROBLEM

Our first task must be to define the problem and then redefine the problem as our knowledge and understanding improves and our experience of the changing climate increases. This is a continuous process.

You should know that if you change all your light bulbs to low energy light bulbs, if you cut your car usage by 50%, if we close the coal burning power stations that we still use to produce nearly half our electricity in this country: if we do all those things it will have no effect whatsoever on climate change. It will not affect it tomorrow; it will not affect it next week, next year, or in ten years' time. If we do all these things we will probably have an effect on the climate in 100 to 200 years time. But you need to be absolutely clear that we will only have that effect if everybody else in the world makes the same sort of cuts that I've just described.

The UK emits about 150 million tons of CO_2 equivalent every year: that is 2% of the CO_2 emitted worldwide. If the UK stopped producing carbon dioxide - which of course is impossible because to do so we would have to switch off all the lights, all the power stations, lock our cars away and turn off the central heating - we would reduce global emissions by 2%. It won't make a great deal of difference. We have to recognize that the United States emits 21% of the world's carbon dioxide while China emits 15%. And China is the world's fastest growing economy and will help to drive world energy consumption to double within less than ten years. China, by the way, has little or no oil and gas, but lots and lots of coal. China's expansion leads to smoke, smog and soot, as well as increasing emissions of CO_2.

RECOGNISE PRIORITIES

I am not suggesting that global warming is not happening. I'm not denying that the greenhouse gases caused by industrialization are part of the problem. I am not suggesting that we do nothing at all. The most important thing though, is to recognize how much and how little we can do. We must look at the priorities and the things that are really important when it comes to living our lives, going to work and indeed to protecting the planet for ourselves and our children and our grandchildren.

Our immediate problem is not global warming or, at least, we can do nothing to stop the global warming that is already in process. There is no silver bullet; there is no way of switching off the global warming process. We can do things which might stop it getting worse in the distant future but we and our children and our grandchildren have to live with the consequences of 150 years of industrialization.

THE ENERGY ISSUE

The immediate threat to our way of life is the coming energy shortage. This could be particularly acute in the United Kingdom as we have become used to being in control of our own energy resources.

We exploited our vast reserves of coal, we found huge oil wells beneath the North Sea and we found gas there as well. The coal industry has been dramatically reduced and the more than 50% of the coal we consume is now imported from abroad, from as far away as Russia, South Africa even China and Vietnam. Our oil reserves are beginning to run out, production levels are falling and we're close to the point where we'll have to start importing oil again.

The biggest source of energy that we use in the United Kingdom is gas and the natural gas in the North Sea is declining to the extent that we already have to import 10% of the gas we require and we expect to be importing 80% within ten years. Within a generation, we have gone from a situation where our energy supplies were under our own control to a situation where we rely on foreign companies and global markets to provide what we need. We've reached this situation at the point when global demand is beginning to outstrip supply.

Peak Oil and Peak Gas are the points at which the reserves reach the maximum output. Beyond these peaks the output of oil and gas on a global scale will slowly decline.

If western countries continue to increase their demand for energy, and developing countries, particularly India and China, increase energy demand at a dramatic rate, the collision of ballooning demand and collapsing supply can only lead to what some people have described as a price spike. Unfortunately this is not an accurate description because a spike goes up and then goes down again. Energy prices will go up but as long as the shortages persist - and we are talking about reaching the end of global resources - then there will be no downward motion.

The popular wisdom is that there are always scare stories and that they are always proved to be wrong. We did not die of SARS, and the Millennium Bug did not destroy our computers. But we're talking here about natural resources, and for us to enjoy an improving lifestyle and to share that with nations now undeveloped implies that these resources will never run out. We all know that this cannot really be true, but we prefer to believe that running out is far, far, far away in the future.

We say, of course, that if prices go up we will exploit oil and gas in more difficult, remote and hostile regions. To some extent that is true. We say that, of course, we should be relying on wind power or hydrogen or biofuels or solar panels or the tides. All of these things have been reviewed and all of them are found wanting. Certainly as far as short term solutions are concerned they offer no answer. It takes years, not weeks, to install a wind-farm. The same goes for solar panel arrays, for tidal barrages, for converting power stations to consume biofuels.

We can extract oil or gas from coal seams. We can extract oil from shale or tar sands, but all these processes require vast amounts of energy. They require vast amounts of water in many cases; of energy, of infrastructure and of skilled people. It will take time and money to bring these people together and even though oil and gas prices may rise out of control, if you look at the value of the energy created against the energy put in to convert all these materials it just doesn't balance. Many of these processes involve serious pollution and waste management concerns as well.

The wind and the sun and the tides are free, say many people. True, but the capacity of the wind-farms that could be built in the UK, of the solar panels that could be installed in the UK, of the tidal barrages and the wave machines that could be put in place is such that doubling, trebling, quadrupling energy from these sources would provide less than 25% of our electricity.

It's not just electricity that we use. Gas is the major fuel. Electricity could be used instead for some applications, but gas is much more efficient since it is converted into energy at the point of use, whereas electricity is transported miles from power stations and significant proportions are lost en route. Local generation is a partial solution, but micro-generation at the domestic level is not cost-effective.

Oil is a major part of our energy equation. It is also a major raw material for plastics, pharmaceuticals and chemicals. Oil is used for transport: for our cars, lorries, buses, trains and aircraft. Generating electricity from solar panels, wind or tides provides nothing that is of any use for driving a transport fleet. Electric vehicles are possible and are much improved with recent battery redesign, but their range and performance still does not match the petrol car. The convenience is not comparable either. A petrol car can be refuelled for 500 miles in five minutes. An electric car needs six hours to recharge for 150 miles if you're lucky.

THE END OF LIFE AS WE KNOW IT

The conclusion of all this is that we are getting to a point where we are going to be unable to continue with a lifestyle that we have become accustomed to. If we put all our emphasis on searching for alternative fuels and alternative sources of energy we are simply going to postpone the inevitable. We are going to be living on false hopes and we are going to suffer a greater disappointment when the crash comes.

The only solution to the coming problems is to reduce our energy consumption dramatically. That, incidentally, will have a significant effect on our CO_2 emissions as well. Reducing energy though is not only a question of turning down the central heating using low energy light bulbs and driving more slowly to improve petrol consumption.

Energy, as we have seen, is part of the process of the creation of everything we eat, use or wear. To reduce our energy consumption we have to reduce our general consumption. The consequences of that, of course, are that if there is less production there will be less employment and there will be less wealth. For those of us in the industrialised West the world will become a less comfortable place to live.

AS CONSUMERS

As consumers we need to seek the best value for energy. To get the very best from every gallon, therm or kilowatt-hour that we buy.

Saving energy also means saving products. Every piece of packaging or junk mail that we throw away means that we have wasted the resources that have been mined, refined and processed to create them and the energy used in production, distribution and then in recycling.

As consumers we will always need energy and it is up to us to choose clean energy. As we have seen in the earlier parts of this book, renewables and biofuels can only satisfy part of the demand. Nuclear power must represent a significant part of our electricity generation capacity if we are serious about reducing carbon dioxide emissions. We need to make a responsible choice.

IN BUSINESS

In business we need to seek the best value in energy for ourselves and our organizations. We need to seek the best value in energy for our clients and to recognise that both our clients and our suppliers will face problems as energy becomes more scarce and more expensive.

Our businesses will prosper if we can develop solutions to overcome these problems for our clients. We will need to find solutions that will allow us to continue to provide a level of service in the face of the difficulties encountered by our suppliers.

AS CITIZENS OF THE WORLD

If Britain cut all its carbon dioxide emissions by cutting out all fossil fuel use, that would reduce global emissions by only 2%. We could say that that is an insignificant amount and therefore we should not bother. But at the very least we must set an example for the rest of the world, because, without the rest of the world, the hope of sensible energy management and carbon dioxide control is fruitless.

As citizens we must take a global view and encourage our government not to try and protect scarce resources by military intervention but to encourage other nations to use these scarce resources wisely and cleanly. To help them develop new generation methods and new technologies so that they can raise their standards of living towards our own with minimum damage to the planet.

We must also accept that climate change is a fact and that the process, whatever the underlying causes, is in motion and cannot be switched off overnight, next week or next year.

We know that the likely consequences are drought, fires and famine. Our nation must help the poorer nations and those most vulnerable to the effects of climate change so they can survive the future. Our efforts now will protect the planet for the future. We are protecting ourselves, we are protecting our children and most of all we are protecting the generations to come.

NO QUICK FIX

While global warming is a major threat, it is widely misunderstood and mishandled. Attacking minorities such as those who drive 4x4s is an empty gesture which will achieve nothing, except to alienate people and encourage them either to totally ignore green issues or to become completely hostile to them.

Beware of crazy technological fixes. The Deputy Mayor of London has proposed a Carbon Act to mirror the Clean Air Act of the 1950s. Under her proposal, households would be inspected and penalized for the amount of carbon that they emitted. While the Clean Air Act in the 1950s solved a very local problem – the London Smog, caused by the fumes from coal fires being trapped above the city - global warming is a global problem and any controls on carbon emissions in London will have an infinitesimal effect on global warming on their own.

Beware as well of grandiose schemes like the 'reflector in the sky'. Some propose sending a reflector into orbit designed to reflect back some of the sun's rays and therefore reduce its heating effect on the earth.

Then there's the plan to launch a fleet of ships to crisscross the oceans day and night, pouring iron compounds into the seas to stimulate the algae in the hope that they will absorb carbon dioxide. Which countries would control such a project? What unforeseen side effects might occur? And do we expect to carry on this process for ever?

Is global warming the greatest threat facing human civilization? The true threat is in governments and organizations overreacting, reacting in ways which will not solve the problem and ignoring the greatest threats: energy shortage and social upheaval.

INTERESTING TIMES

We live in interesting times. We must constantly be aware of those times and adapt our actions and behaviour as the future unfolds. There's no magic bullet. Only a moving target. And the human race is caught in the crossfire.

Changing Times

Every day something new about the climate is reported. Since this book went to press the IPCC has issued its Fourth Assessment Report (FAR) and the papers have been full of comment. Every day governments announce new initiatives while scientists claim that no-one is doing enough. On the other side of the argument, there are still plenty of people who are adamant that global warming is nothing but a hoax.

For the up-to-date position, background information, debate and further reading visit;

www.willclimatechangeyourlife.com

Further copies of this book can be order from the publishers at;

www.ecademy-press.com

About the Author

Anthony Day – speaker, writer and management accountant - is not a member of Greenpeace, the Greens, Friends of the Earth or any campaigning group. His views are his own. As an accountant he is concerned with the practical effects of climate change and energy on business, society and our fundamental economic survival. He has addressed management audiences throughout Europe and advises corporates on climate and energy issues. He is a Fellow of the Chartered Institute of Management Accountants and a Fellow of the Institute of Sales and Marketing Management.

His next book *"Will climate change your business?"* *(Autumn 2007)* looks at corporate social responsibility in the context of climate change and the coming energy crisis.

Printed in the United States
107884LV00001B/101/A